Collins

KS3 Revision Science

Science

KS3

Revision Guide

Byron Dawson and
Eliot Attridge

Contents

0 — 1000ml
APPROX.
100 — 900
800
700
Ex20°C
100ml ±1 ml
0 100
10 90
20 80
30 70
40 60
50
40
30
20
10

Contents

Biology

Cells – the Building Blocks of Life

You must be able to:

- Use a microscope to help understand the functions of the cell
- Remember the differences between animal and plant cells
- Understand how substances move into and out of cells by diffusion
- Understand the organisation of cells.

Using a Light Microscope

- Cells are too small to see with the naked eye. Using a light microscope helps us to see and draw cells.

A plant cell drawn after observation with a light microscope

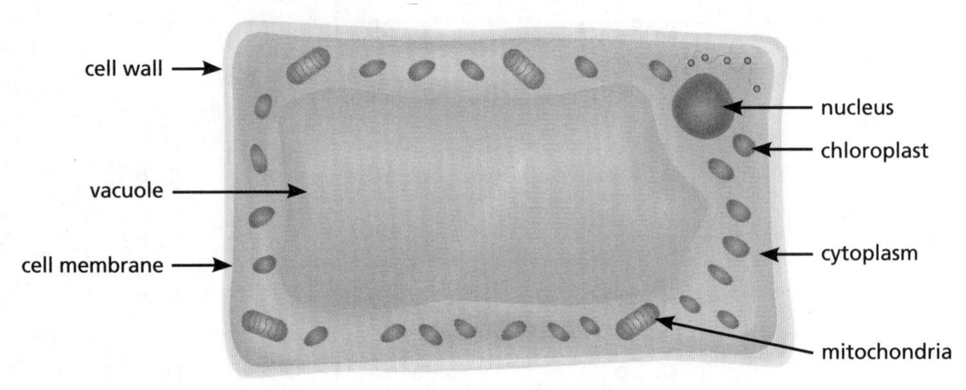

Key Point

Objects need to be placed on a slide, stained and covered with a coverslip, placed on the 'stage' of the microscope, illuminated and then focussed.

How Plant and Animal Cells Work

- Animal and plant cells share some features but not others.
- Different parts of animal and plant cells have different functions.

Part	Function	Animal Cells?	Plant Cells?
Membrane	Controls what enters and leaves the cell	Yes	Yes
Cytoplasm	Place where lots of chemical reactions (photosynthesis) take place	Yes	Yes
Nucleus	Stores information (in DNA) and controls what happens in the cell	Yes	Yes
Mitochondria	Release energy from food (glucose) by aerobic respiration	Yes	Yes
Cell wall	Made from cellulose and gives rigid support to the cell	No	Yes
Vacuole	Inflates the cell like air pumped into a tyre and provides support to the cell	No	Usually
Chloroplast	Contains green chlorophyll that changes sunlight energy into glucose food energy	No	Yes

Diffusion

- **Diffusion** is one of the ways that substances enter and leave cells.
- In an animal cell, oxygen and glucose diffuse through the membrane into the cell. This is because there is more oxygen and glucose outside the cell than there is inside.
- Carbon dioxide and waste products diffuse out of the cell into the blood.
- In a plant cell, carbon dioxide diffuses in. Oxygen and glucose diffuse out.

Key Point

Diffusing substances always move from where there is a lot of the substance (high concentration) to where there is very little (low concentration).

Unicellular Organisms

- **Unicellular** organisms have just one cell.
- *Euglena* has a long whip-like structure to help it move through water.
- *Amoeba* can make finger-like projections to catch food.

Amoeba as seen through a microscope

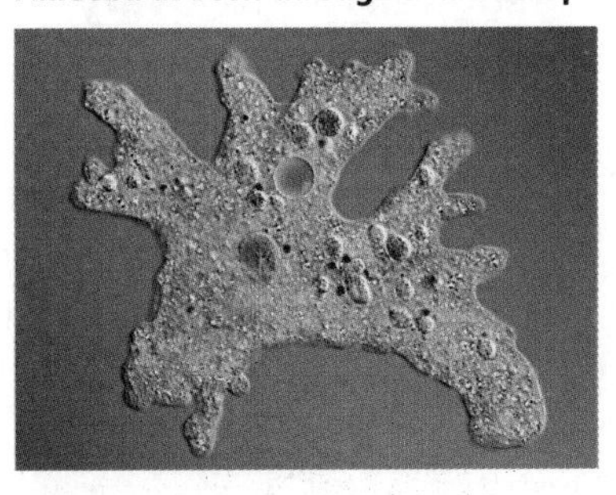

Organisation of Cells

- Cells of the same type carrying out the same function are usually grouped together to form a **tissue**, e.g. skin cells.
- Different types of tissue are grouped together to form **organs**, e.g. the brain.
- Different types of organs are grouped together to form **organ systems**, e.g. the nervous system.
- Different types of organ systems work together to form the organism, e.g. a human being.
- Examples of cell and organ systems include:
  - Bone cells in the skeletal system
  - Blood cells in the transport system
  - Nerve cells in the nervous system
  - Sperm cells in the reproductive system.

Key Point

cells ➠ tissues ➠ organs ➠ systems ➠ organisms

Quick Test

1. Name one structure that is found in plant cells but not animal cells.
2. Where in a cell is energy released from food?
3. Name the process where molecules move from where there are lots of them to where there are only a few.
4. Put these words in order of complexity starting with 'cell': cell, organism, organ, system, tissue.

Key Words

membrane
cytoplasm
nucleus
mitochondria
cell wall
vacuole
chloroplast
diffusion
unicellular
tissue
organ
organ system

Biology

Cells – the Building Blocks of Life

You must be able to:

- Understand and explain the structure of the human reproductive system and how it works
- Know how reproduction and fruit dispersal works in a flowering plant
- Understand why plant reproduction is important to humans.

Reproduction in Humans

- Sexual reproduction in humans involves males and females. Males produce **sperm** cells in the **testes**. Females produce **egg cells** in the **ovary**.
- The penis deposits the sperm in the female vagina.
- Sperm swim up through the **uterus** to the oviduct.
- **Fertilisation** occurs when a sperm cell joins with an egg cell.
- The fertilised egg then grows into an **embryo** and eventually becomes a baby.

Fertilisation in animals

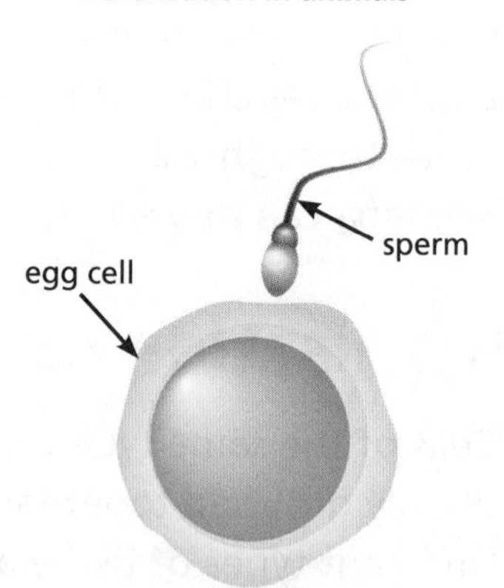

Female reproductive system

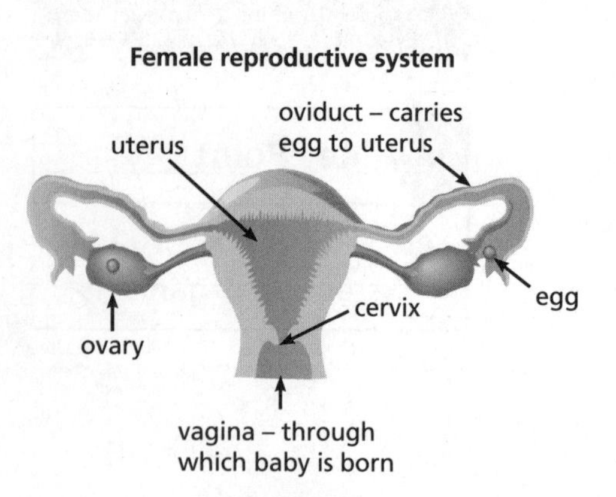

Male reproductive system

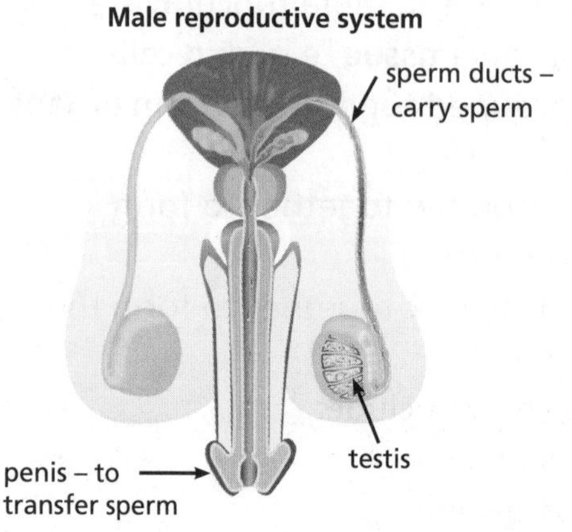

Menstrual Cycle

- Females have a menstrual cycle lasting for about 28 days. This is called **menstruation**.
- On days 1–5, if pregnancy has not occurred, the uterus lining breaks down, tissue and blood are lost, and is replaced with new tissue.
- Fertilisation can only occur on or around day 14 when an egg is released from the ovary.

Gestation

- **Gestation** is the process of the embryo developing in the womb.
- The growing baby receives food and oxygen from the mother's blood through the placenta and umbilical cord.
- Therefore, if the mother smokes and drinks alcohol the baby will also receive some of the alcohol and nicotine.
- In humans, gestation ends after nine months with the birth of the baby.

Key Point

A human foetus takes 38 weeks to grow from the cell being fertilised to a baby.

Reproduction in Flowering Plants

- Some flowers are insect pollinated, e.g. a rose.
  - Insects visit flowers to collect sweet **nectar**
  - They transfer pollen from the **anther** of one flower to the **stigma** of another flower
  - The male pollen fertilises the female egg cell.
- Some flowers are wind pollinated, e.g. grass.
  - Wind blows pollen from one flower to another
  - Wind pollinated flowers do not have a scent or nectar and petals are not brightly coloured as they do not need to attract insects
  - They have a feathery stigma to catch the pollen.

Key Point

Pollination is the transfer of pollen by insect or wind from the anther of one flower to the stigma of another.

Fertilisation is when male and female sex cells join.

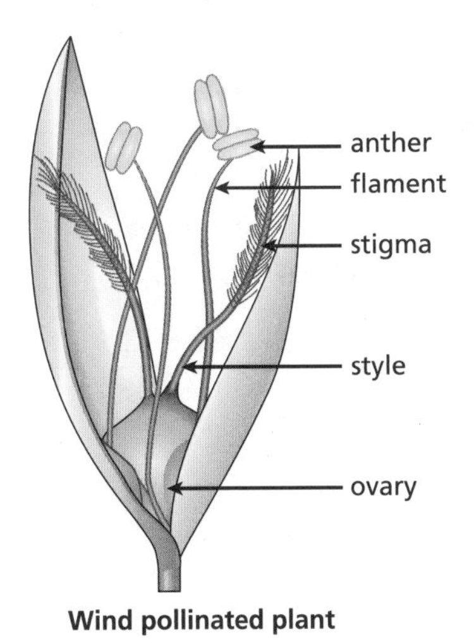

Wind pollinated plant

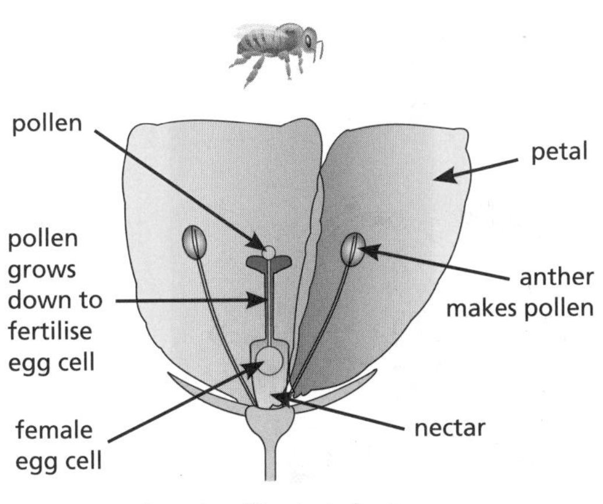

Insect pollinated plant

Dispersal

- After fertilisation, seeds develop inside fruits. These then must be spread over a large area by **dispersal**.
- Some fruit and seeds are spread by animals, e.g. some seeds have hooks which stick to an animal's fur.
- Some are spread by wind. These often have wings or parachutes to be carried by the breeze, e.g. sycamore and dandelion seeds.
- Plants produce many seeds as most fail to grow into a new plant.

The Importance of Plant Reproduction

- Plants provide us with most of our food.
- Without insects to pollinate the flowers, many of us would starve due to lack of food.

Quick Test

1. Which two cells join together at fertilisation?
2. On which day of the menstrual cycle is a female egg released?
3. Write down the differences between an insect pollinated flower and a wind pollinated flower.
4. List two ways that fruits and seeds can be dispersed.

Key Words

sperm
testes
egg cell
ovary
uterus
fertilisation
embryo
menstruation
gestation
nectar
anther
stigma
pollination
dispersal

Biology

Eating, Drinking and Breathing

You must be able to:

- Know and explain how humans move air into and out of lungs
- Know and understand how oxygen and carbon dioxide move between the blood and the lungs
- Understand the effect of exercise, asthma and smoking on the breathing systems.

Breathing

- Breathing involves moving air into and out of the lungs.

When breathing in:

1. Ribs move up and out
2. **Diaphragm** flattens and moves down
3. Space inside the lungs increases
4. This increases the volume and reduces the pressure
5. Air rushes into the lungs from outside.

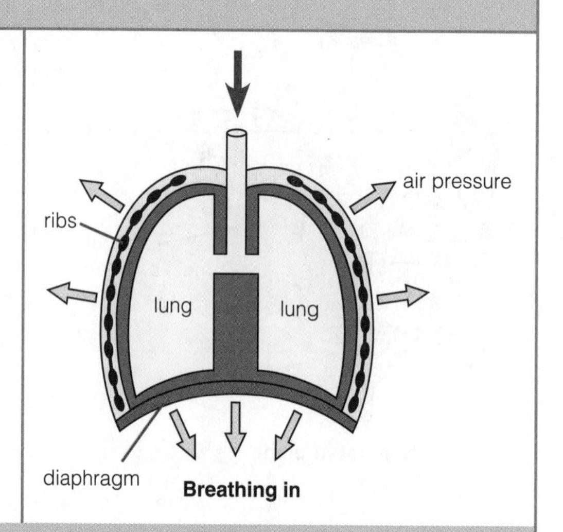

When breathing out:

1. Ribs move down and in
2. Diaphragm moves up
3. Space inside the lungs decreases
4. This decreases the volume and increases the pressure
5. Air is pushed out of the lungs.

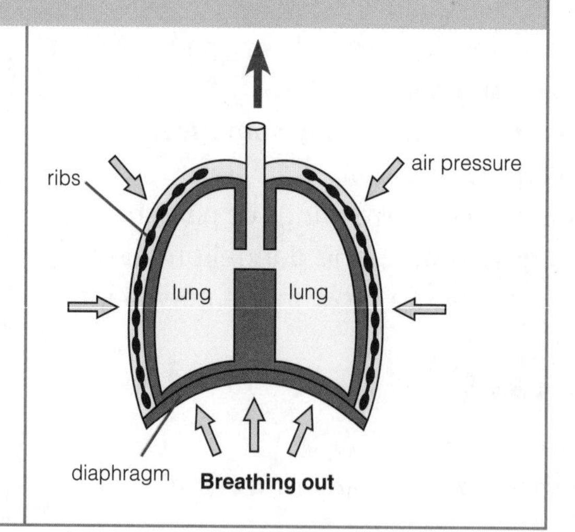

Key Point

When we breathe in, air is pushed in by **air pressure** from the outside.

Gas Exchange

- The lungs are made of millions of tiny air sacs called **alveoli**.
- These air sacs are:
  - Thin
  - Moist
  - Have a good blood supply
  - Have a large surface area.

Structure of alveoli

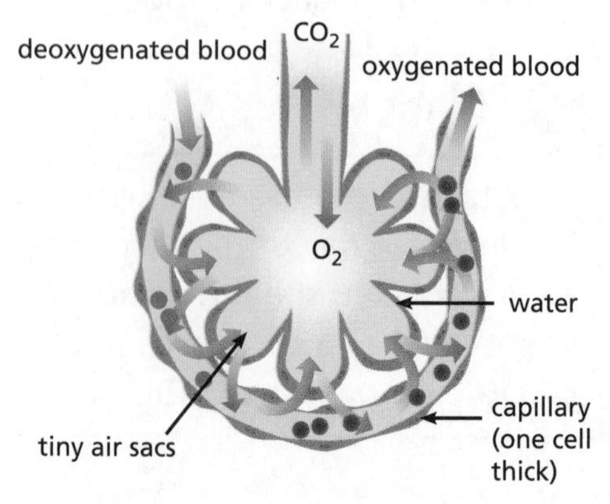

- Gas exchange is when:
  - Carbon dioxide leaves the blood and enters the lungs to be breathed out
  - Oxygen leaves the lungs and enters the blood.
- Gas exchange happens through the thin walls of the air sacs.
- The exchange happens because of diffusion (see page 4).

Key Point

Diffusion is the movement of particles from a high to a low concentration.

The breathing system

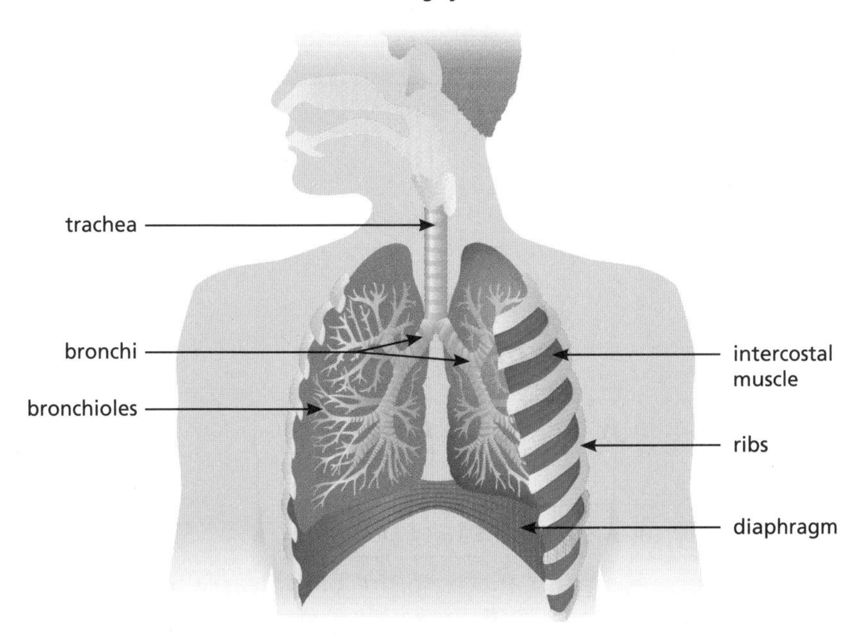

Things that Affect our Breathing

Things that Affect Breathing	Effect on Breathing
Exercise	• Increases lung size. • Improves gas exchange.
Asthma	• Causes breathing tubes (bronchioles) to narrow, making breathing difficult.
Smoking	• Damages the breathing tubes so that mucus builds up. This causes a cough, makes breathing more difficult and makes infections more likely. • In the long term can cause emphysema and lung cancer.

Quick Test

1. Explain how the ribs and diaphragm move to make you breathe in.
2. Explain what happens to the volume and pressure inside the lungs when the ribs move down and in.
3. Name the process by which oxygen moves from the air in our lungs into the blood.
4. Describe the effect smoking has on the lungs.

Key Words

diaphragm
air pressure
alveoli
asthma

Biology

Eating, Drinking and Breathing

You must be able to:
- Explain what is meant by a healthy diet
- Explain the energy content of a healthy diet and understand what happens when a healthy diet becomes unbalanced
- Know and explain the jobs of different parts of the digestive system.

A Healthy Diet

- A healthy diet contains all the right proportions of the following substances:

Content of Healthy Diet	Purpose
Carbohydrate	Gives the body energy
Fat	Gives the body energy and can be stored in the body
Protein	Used for growth
Vitamins	Used to help chemical reactions take place in the body
Minerals	Used to make bones strong and help the blood carry oxygen
Fibre	Helps undigested food pass quickly through the gut
Water	Dissolves chemicals so that chemical reactions can take place

- A healthy diet also contains sufficient food to provide us with just the right amount of energy.
- Energy in food is measured in calories or joules.
- 1 calorie = 4.2 joules
- A young man needs about 2500 kcal per day.
- 2500 kcal × 4.2 joules = 10,500 kJ per day.

Key Point

When dieticians talk about calories in food they really mean kilocalories. A kilocalorie is 1000 calories.

An Unbalanced Diet

- Eating an unbalanced diet can cause many problems:

Cause	Problem
Eating too much	Obesity
Eating too little	Starvation/malnutrition
Not eating enough protein	**Kwashiorkor**, an illness caused by severe protein deficiency. It is mostly seen in developing countries.
Not eating enough vitamins	A lack (or deficiency) of different vitamins causes different diseases, e.g. a lack of vitamin C causes **scurvy**.
Not eating enough minerals	A lack of iron causes **anaemia**. A lack of calcium causes soft bones.

The Digestive System

- The digestive system processes food that is eaten in the mouth. Food travels through the **oesophagus**, **stomach**, **intestine** and **rectum** until the waste is eliminated from the **anus**.
- The **pancreas** also plays a key role in digestion by producing digestive enzymes which help break down the food.

The digestive system

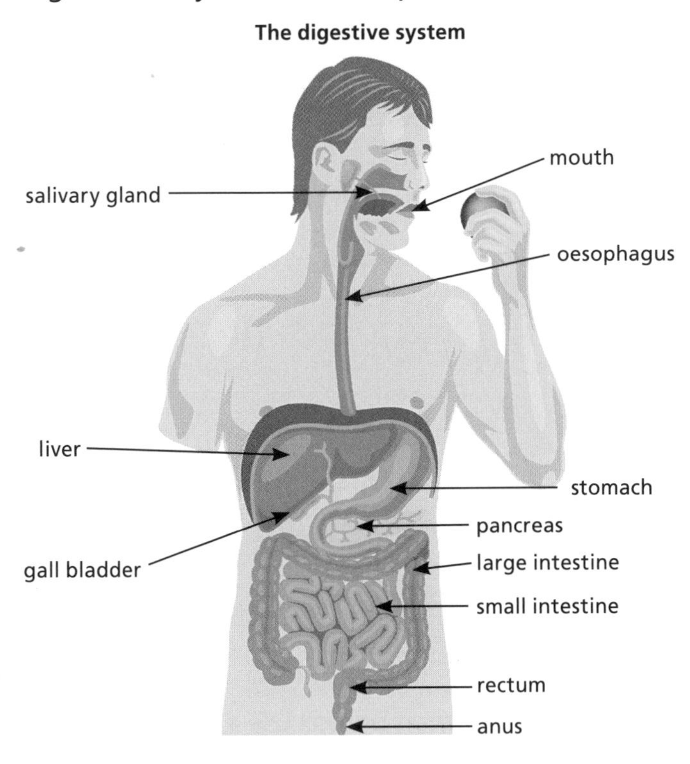

Food in Plants

- Unlike animals that eat food, plants make their own food.
- The process is called **photosynthesis**.
- Plants take water and minerals from the soil.
- They take carbon dioxide from the air.
- They use energy from the Sun to convert these substances into carbohydrates in their leaves:

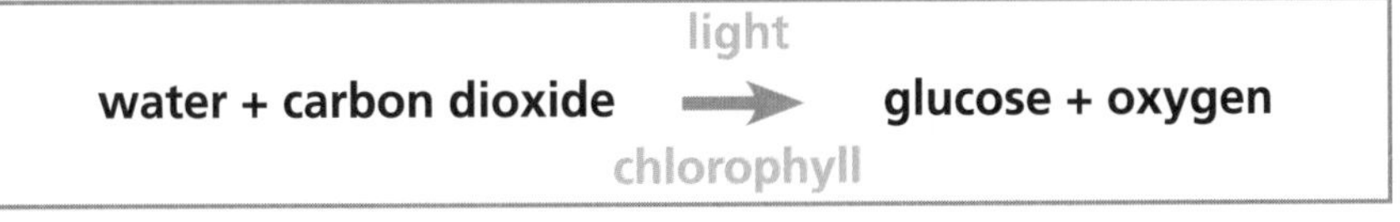

Quick Test

1. Name five components of a healthy diet.
2. Name three possible consequences of eating an unbalanced diet.
3. Write down the different parts of the digestive system in the order food travels through them. Start with **mouth**.
4. Describe the difference in feeding between plants and animals.

Key Point

Animals eat food, plants make it.

Key Words

carbohydrate
fat
protein
vitamins
minerals
fibre
kwashiorkor
scurvy
anaemia
oesophagus
stomach
intestine
rectum
anus
pancreas
photosynthesis

Review Questions

Key Stage 2 Concepts

1 All living organisms have certain things in common.

a) Copy the table below and put a tick (✔) in the box next to the characteristics found in all living things.

Hardness	
Nutrition	
Transparent	
Flying	
Flexible	
Growth	
Reproduction	
Melting	

[3]

b) One of the characteristics of living things is movement. Human beings use a skeleton to help them move. The skeleton also protects different parts of the body.

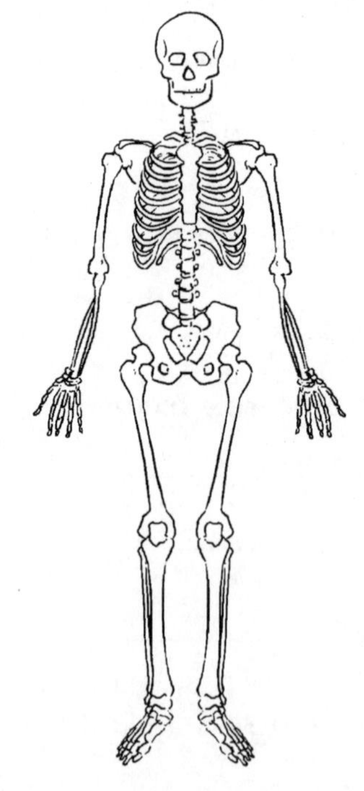

Trace or sketch the diagram of the skeleton.

i) Draw a **J** on the diagram to show the position of a joint for movement. [1]

ii) Draw a **P** on the diagram to show part of the skeleton that protects the body. [1]

2 Materials have many different properties. Jack found these materials in his father's shed.

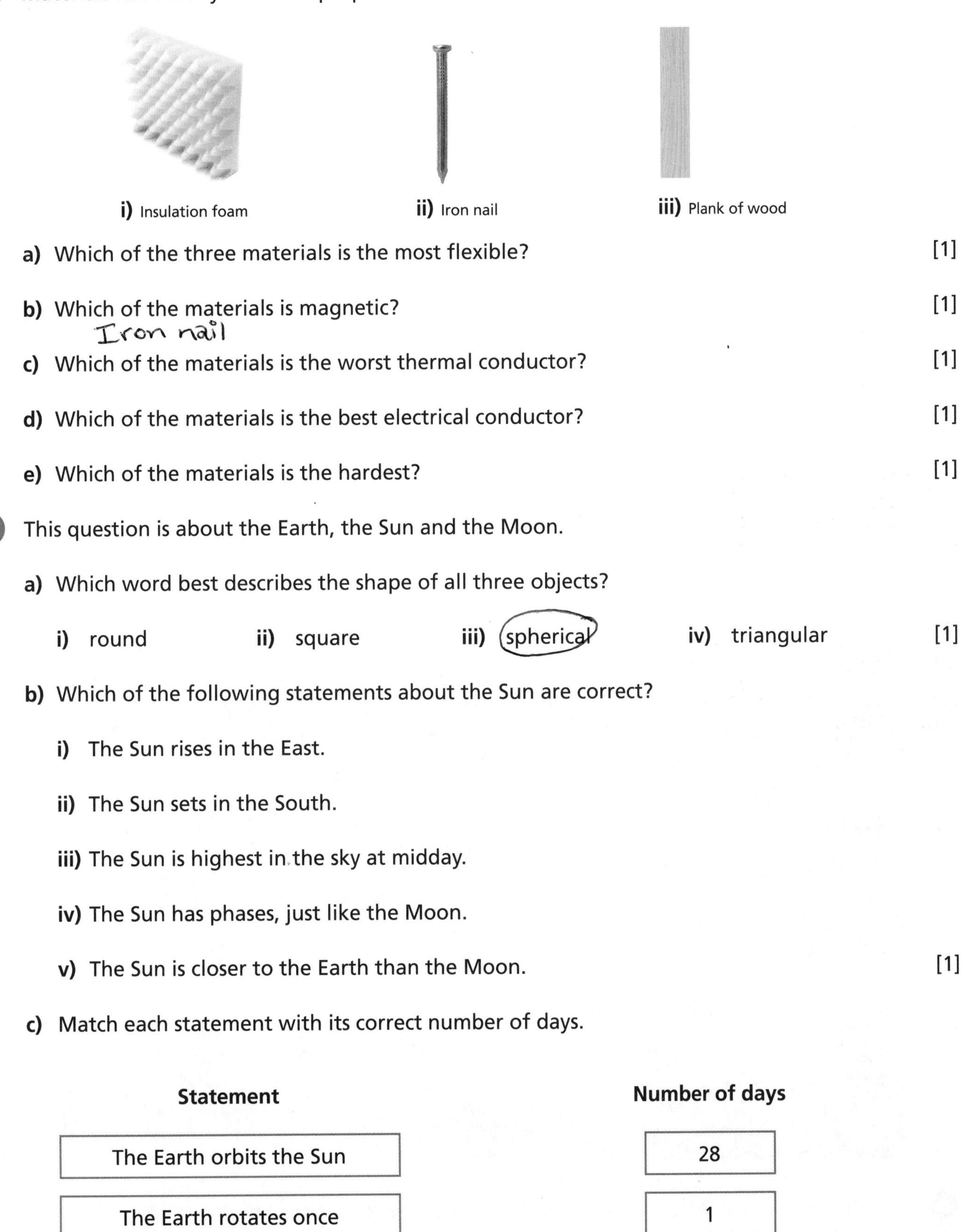

i) Insulation foam **ii)** Iron nail **iii)** Plank of wood

a) Which of the three materials is the most flexible? [1]

b) Which of the materials is magnetic? [1]

Iron nail

c) Which of the materials is the worst thermal conductor? [1]

d) Which of the materials is the best electrical conductor? [1]

e) Which of the materials is the hardest? [1]

3 This question is about the Earth, the Sun and the Moon.

a) Which word best describes the shape of all three objects?

i) round **ii)** square **iii)** spherical **iv)** triangular [1]

b) Which of the following statements about the Sun are correct?

i) The Sun rises in the East.

ii) The Sun sets in the South.

iii) The Sun is highest in the sky at midday.

iv) The Sun has phases, just like the Moon.

v) The Sun is closer to the Earth than the Moon. [1]

c) Match each statement with its correct number of days.

Statement	Number of days
The Earth orbits the Sun	28
The Earth rotates once	1
The Moon orbits the Earth	365

[3]

Cells – the Building Blocks of Life

1 Match the part of a cell to its function.

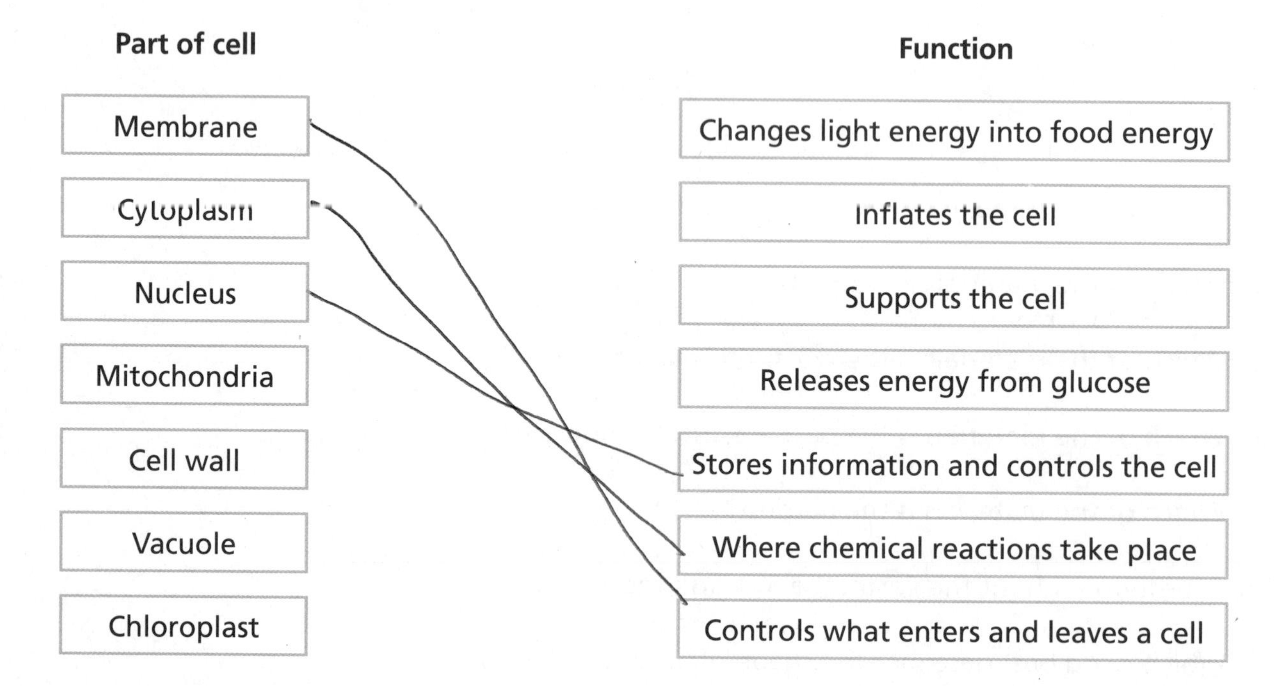

[7]

2 These plant cells were seen using a microscope. Make a labelled drawing of one of them.

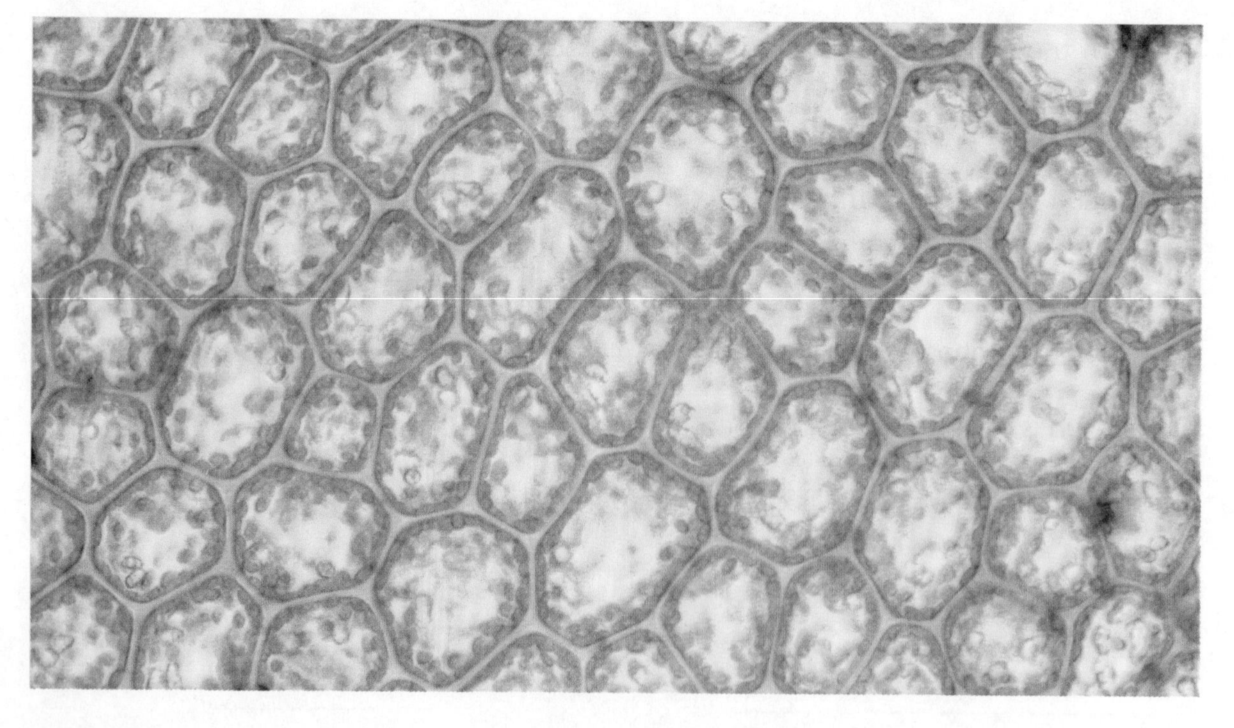

[6]

3 Which of these words describes how substances can enter or leave a cell?

i) cytoplasm **ii)** vacuole **iii)** diffusion [1]

Eating, Drinking and Breathing

1 Humans need to eat a healthy diet.

a) Explain what is meant by a healthy diet. [2]

b) Write down three things that can happen if we do not have a healthy diet.

Explain your three answers. [3]

2 Look at the diagram of the digestive system.

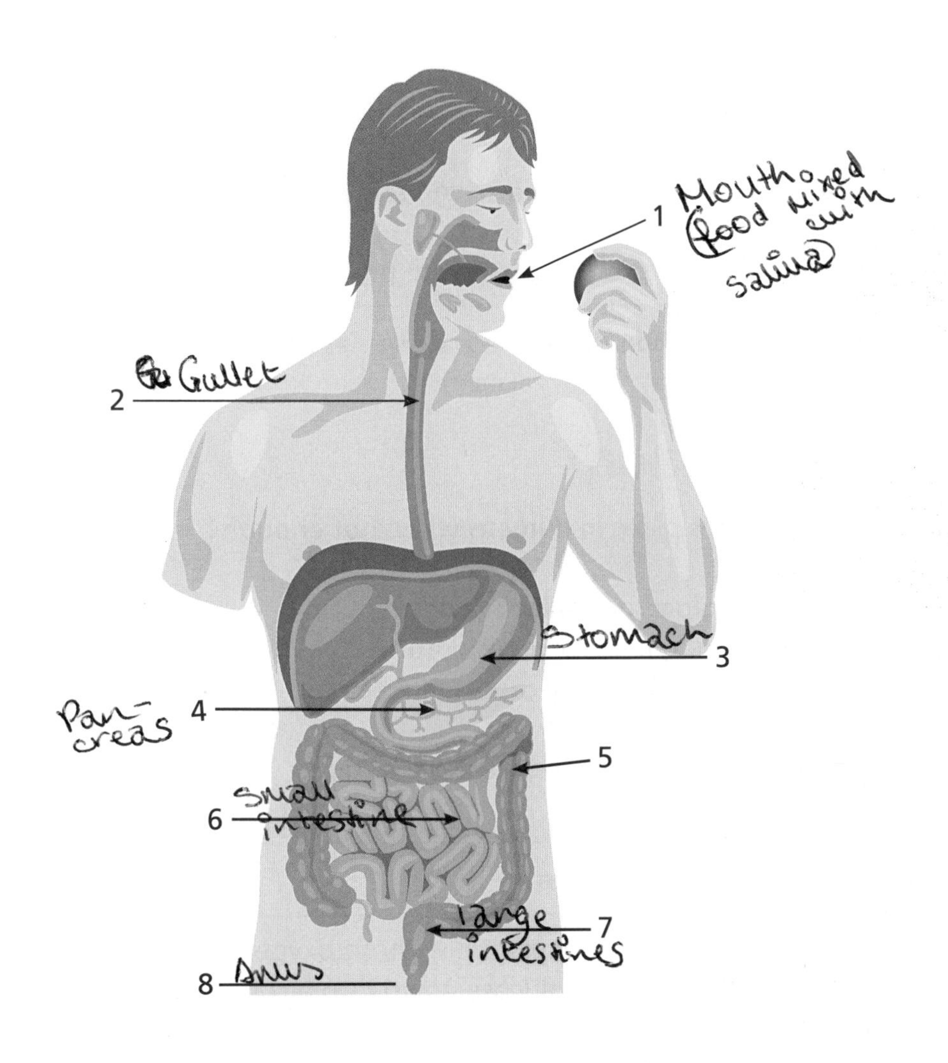

a) Give the correct names for each part 1–8. [8]

b) Explain what each of the parts you have labelled does. [8]

3 Explain the difference between feeding in animals and feeding in plants. [4]

Getting the Energy your Body Needs

You must be able to:
- Explain what respiration is
- Understand and explain aerobic and anaerobic respiration, including the differences between them.

Respiration

- **Respiration** is the process by which organisms release energy from food.
- The energy is needed to power all the chemical processes necessary for life.
- There are two types of respiration, **aerobic** and **anaerobic**.

Aerobic Respiration

- Humans release energy from **glucose** and **oxygen** by aerobic respiration.
- Carbon dioxide and water are produced as waste products.

glucose + oxygen → carbon dioxide + water + energy

Anaerobic Respiration

- Anaerobic respiration takes place in humans when not enough oxygen is available.
- Humans can break down glucose into **lactic acid**.
- Less energy is released during anaerobic respiration.
- Lactic acid is also released. This quickly causes muscle pain and fatigue.
- 'Getting the burn' is when muscles produce lactic acid in anaerobic respiration.

glucose → lactic acid + energy

- Yeast is a microorganism that can also respire without oxygen (anaerobic respiration). Yeast breaks glucose down into alcohol and carbon dioxide.
- This process is called **fermentation**.

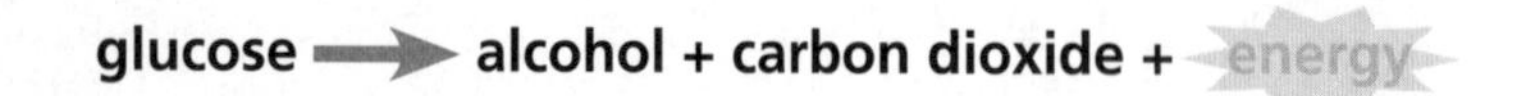

glucose → alcohol + carbon dioxide + energy

Key Point

Although humans can respire without oxygen we can only do this for a very short time. It happens when we need a lot of oxygen very quickly, such as when we run a fast race.

Key Point

Fermentation is used to produce alcoholic drinks such as wine and beer.

Similarities and Differences between Aerobic and Anaerobic Respiration

	Aerobic	Anaerobic
Use glucose	✔	✔
Use oxygen	✔	✗
Produce carbon dioxide	✔	✔ Fermentation in yeast ✗ but not in humans
Produce water	✔	✗
Release *lots* of energy	✔	✗
Can produce lactic acid	✗	✔ In humans ✗ but not by fermentation
Can produce alcohol	✗	✔ Fermentation in yeast ✗ but not in humans
Causes muscle fatigue	✗	✔

Quick Test

1. What is the type of respiration that uses oxygen?
2. Name the type of respiration that releases the most energy.
3. Give the type of respiration that can produce lactic acid.
4. Which type of respiration occurs during fermentation?
5. Other than carbon dioxide, what substance is produced during fermentation?

Key Point

Both aerobic and anaerobic respiration release energy from glucose but aerobic respiration is more efficient and releases more energy.

Key Words

respiration
aerobic
anaerobic
glucose
oxygen
lactic acid
fermentation

Getting the Energy your Body Needs

You must be able to:

- Explain the structure and function of the human skeleton
- Explain how muscles make the skeleton move.

The Human Skeleton

- The human skeleton has several different functions:
  - Supports the body and gives it shape.
  - Acts as a framework that enables muscles to move the body.
  - Protects parts of the body, for example, the skull protects the brain and the ribs protect the heart and lungs.
  - Makes red blood cells in the marrow of the long bones, for example, humerus and femur.

Human skeleton

skull
ribs
humerus
spine
elbow joint
ulna and radius
hip joint
femur
kneecap

Bone marrow

Joints and Muscles

- Bones in the skeleton are held together in **joints**.
- Joints allow the skeleton to move.
- Joints are held together by **ligaments**.
- The end of each bone is covered in **cartilage** for a smooth surface that cushions the joint.
- The joint is filled with a fluid that lubricates the joint.

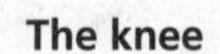

The knee

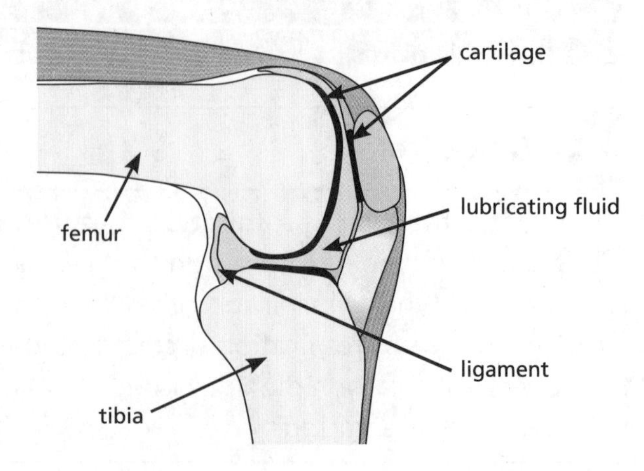

Muscles and Force

- Muscles move joints.
- Muscles are attached to bones by **tendons**.
- Each joint needs two muscles to make it work. This is called an **antagonistic pair**.
- One muscle moves the joint in one direction. The other muscle moves the joint in the opposite direction.
- Muscles work by contracting and getting shorter in length. This pulls the bone and moves the joint.

Muscles in the arm

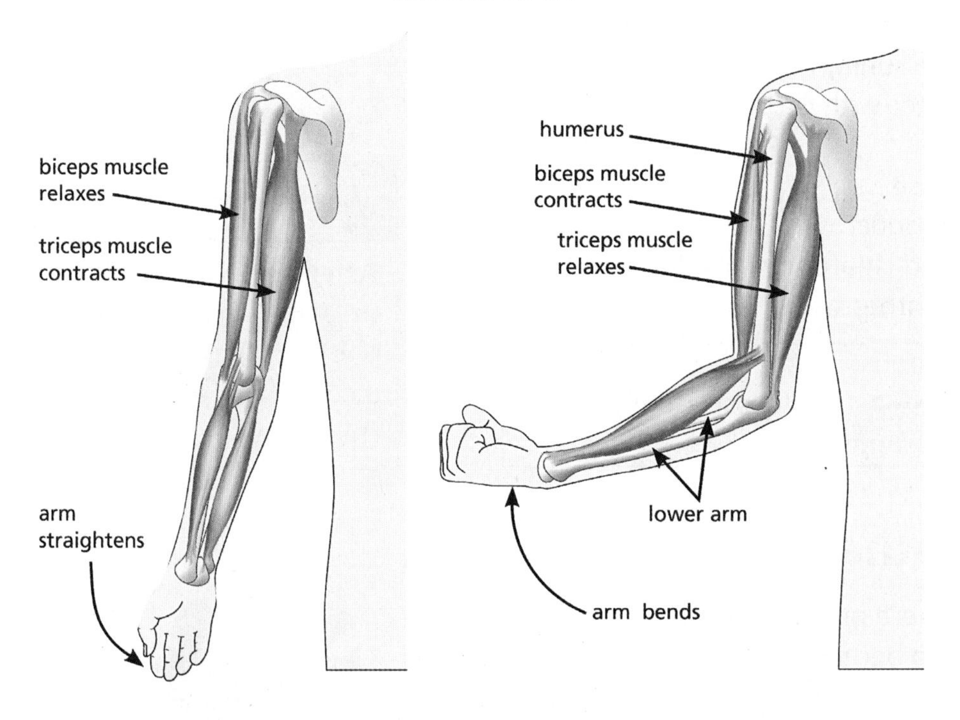

- The force exerted by muscles can be measured in Newtons.
- To work out the force applied by a muscle multiply the mass lifted by its distance from the joint.
- The answer is mass × distance from joint equals force x distance from the same joint. So in the diagram:

10 kg × 30 cm = 300 ? N × 5 cm = 300 therefore the force (?) = 60 N

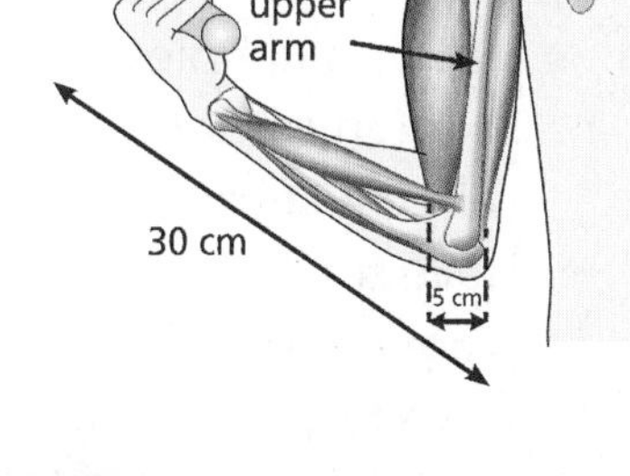

Key Point

Some muscles work in pairs called antagonistic pairs. When one muscle contracts the other muscle relaxes.

Quick Test

1. Give four functions performed by the skeleton.
2. Name the tissue that attaches bones to each other in a joint.
3. Name the tissue that attaches muscle to bone.
4. Explain what antagonistic means.

Key Words

joint
ligament
cartilage
tendon
antagonistic pair

Looking at Plants and Ecosystems

You must be able to:
- Explain how photosynthesis takes place
- Understand how a green leaf is adapted for photosynthesis
- Understand the importance of photosynthesis to other living things.

Photosynthesis

- **Photosynthesis** is the process by which green plants make food.
- Green plants absorb energy from sunlight.
- They use the energy to react water with carbon dioxide to make **glucose**.
- The energy is stored in the glucose.
- **Oxygen** is released as a waste product.
- Plants use a green chemical called **chlorophyll** inside **chloroplasts** to perform photosynthesis.

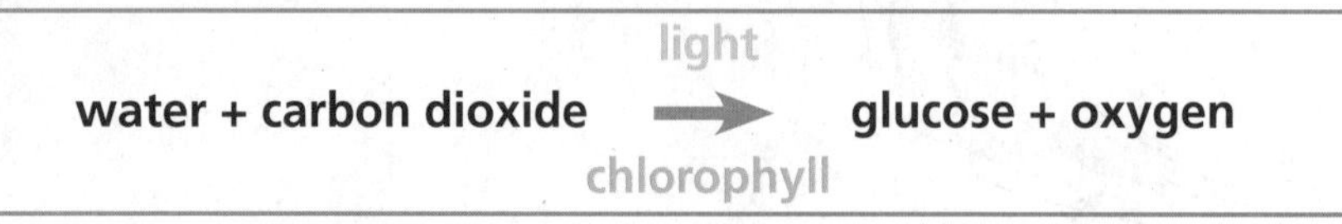

Key Point

When you write down the word equation for photosynthesis always include light and chlorophyll.

Leaves and Photosynthesis

- Leaves are the plant's factory where photosynthesis takes place.
- Leaves are adapted to do this job because they:
  - Are thin – this stops the leaves from being heavy, enabling trees to have more leaves and therefore a larger surface area
  - Have a large surface area – to catch as much sunlight as possible
  - Are green, because of the chemical chlorophyll they use in order to photosynthesise
  - Have small holes called **stomata** on the underside of the leaf to let in carbon dioxide and let out oxygen. A single hole is called a **stoma**
  - Have tiny tubes called xylem to carry water and minerals up from the roots
  - Have tiny tubes called phloem to carry glucose away for storage.

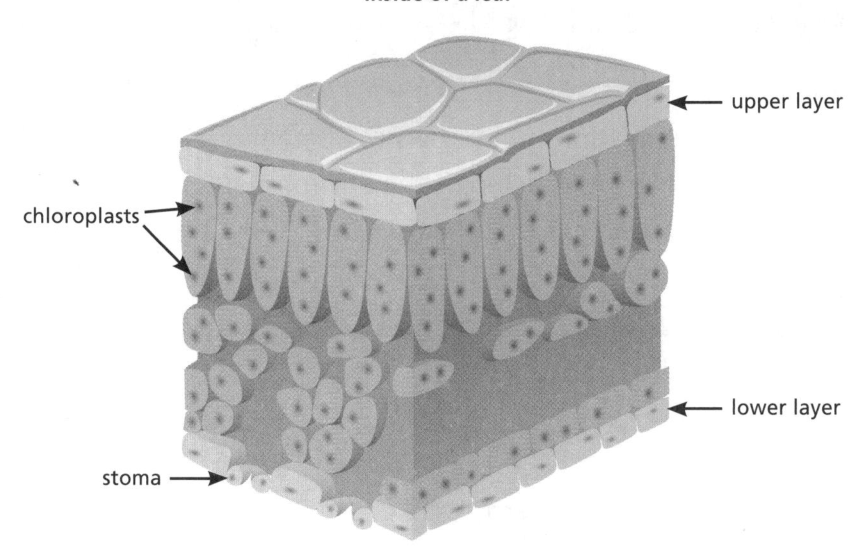

The Importance of Photosynthesis

- Plants and animals depend on each other for survival – they are **interdependent**.
- Plants provide animals with food (glucose) and oxygen.
- Animals provide plants with carbon dioxide that they need for photosynthesis.
- Photosynthesis builds up complex glucose molecules from simple molecules (water and carbon dioxide). This stores energy.
- Respiration breaks down complex glucose molecules into simple molecules and releases energy.

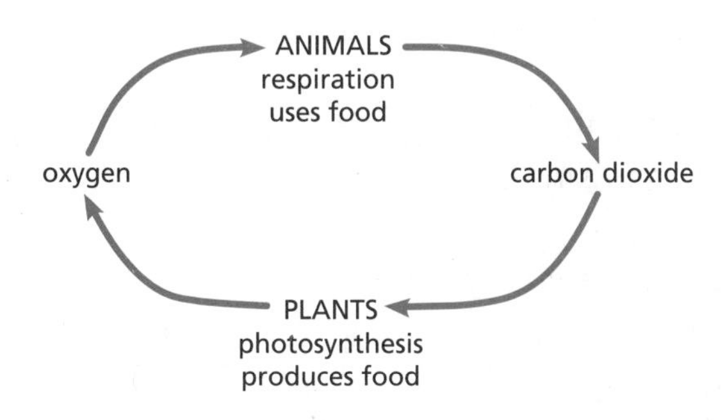

- Most animals on Earth depend on plants for glucose, and oxygen to breathe.
- Most glucose and oxygen are produced by plants in the rain forest and algae in the oceans.

Quick Test

1. Write down the word equation for photosynthesis.
2. State how photosynthesis is different to respiration.
3. Write down three different parts of a leaf and say how each one helps photosynthesis to take place.
4. Explain how plants and animals depend upon one another.
5. Explain what interdependence means.

Key Point

Photosynthesis and respiration are the opposite of each other.

- Respiration uses food and oxygen, and produces carbon dioxide.
- Photosynthesis uses carbon dioxide, and produces food and oxygen.

Key Words

photosynthesis
glucose
oxygen
chlorophyll
chloroplast
stomata
stoma
interdependent

Biology

Looking at Plants and Ecosystems

You must be able to:

- Use food webs to explain relationships between different organisms
- Understand how organisms are affected by the environment
- Understand how differences between organisms help them to survive.

Humans and Their Food Supply

- A good food supply is important for humans.
- This food supply depends on how organisms transfer energy from one to another.
- This means that organisms in an environment are interdependent in many ways.
- Insects **pollinate** flowers so seeds and fruit can grow and be used as food by other animals.
- Humans rely on insect pollinators for many of our crops.
- Insecticides can kill harmful insects and pests, but can also kill useful pollinators.

Interdependence of Organisms

- The best way to show how organisms depend on one another is to draw a **food web**.
- Food webs are made up from many different **food chains**. They are all interdependent.
- A food chain describes what eats what in a community.
- Interdependence describes how all the living organisms in an **environment** depend upon one another.
- Food webs show how organisms depend upon one another for food; they show relationships between organisms.
- **Producers** are plants. They produce food by photosynthesis.
- **Consumers** are animals. They consume food for energy.

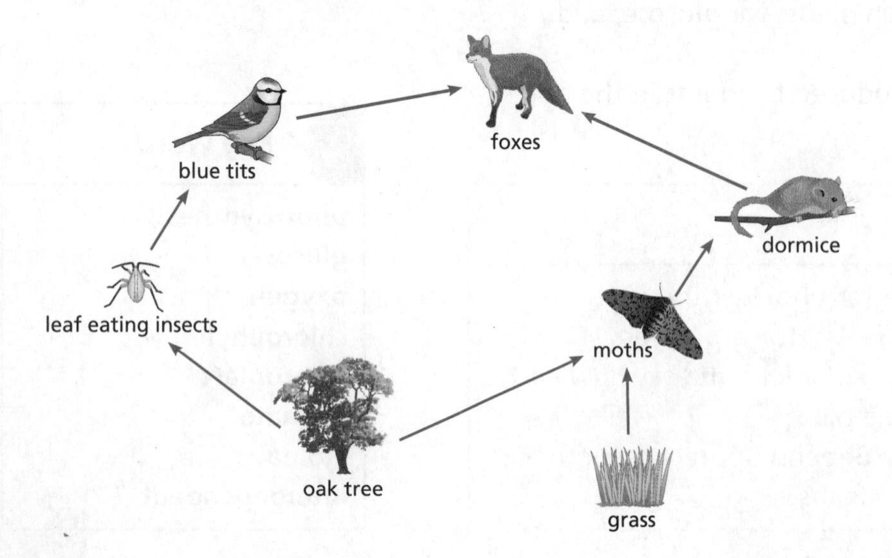

Key Point

The arrows in a food web show how energy is transferred as food from one organism to another.

Organisms and the Environment

- Sometimes poisonous waste can get into an ecosystem:
  1. Plants at the bottom of the food web absorb the poison.
  2. The poison is passed on to the animals that feed upon them.
  3. Because these animals eat lots of plants they absorb more of the poison.
  4. The poison accumulates as it is passed up the food web; this is called **bioaccumulation**.
  5. Eventually there is enough poison in the animals at the top of the food web to kill them:

Poison in a food web

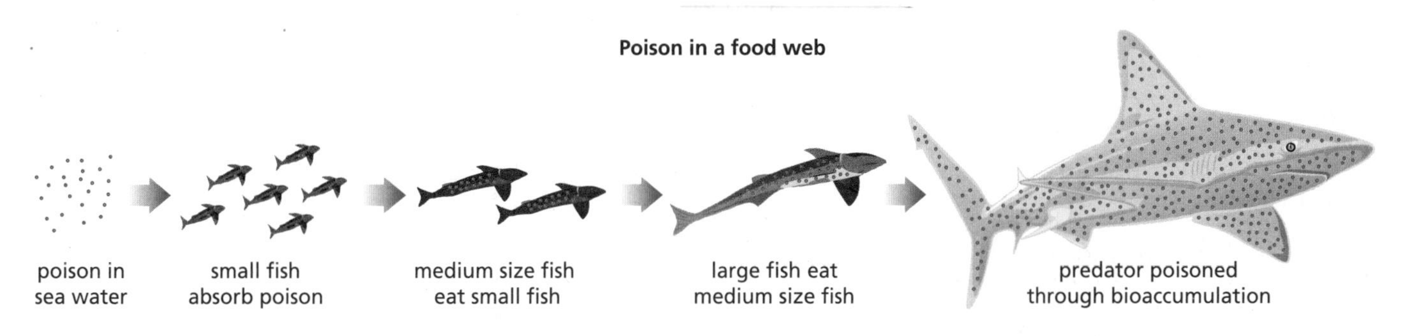

- Bioaccumulation occurs because there are many more organisms at the bottom of a food chain and only a few predators at the top.
- This forms a **pyramid of numbers** that means poisons accumulate in greater numbers in the top predators.

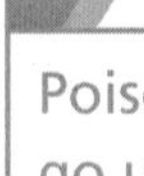

Key Point

Poison builds up as you go up the food chain. This is bioaccumulation.

Variation Between Organisms

- Most living organisms are different from one another. This is called **variation**.
- Different **species** have different characteristics.
- Different species survive in the same ecosystem because they are adapted to survive in different parts of the ecosystem.
- For example, different parts of the ecosystem are called **niches**.
- Fish live in water so they have gills; birds fly, so they have wings.
- Variation means that certain members of a species are more likely to survive when the environment changes.

Quick Test

1. What do arrows on a food web show?
2. What does bioaccumulation mean?
3. Explain why variation between different organisms is important.

Key Words

pollinate
food web
food chain
environment
producers
consumers
bioaccumulation
pyramid of numbers
variation
species
niches

Biology Review Questions

Cells – the Building Blocks of Life

1. Unicellular organisms have different structures from each other.

   Explain why. [2]

2. Match the type of cell with the correct organ system.

Type of Cell	Organ system
Bone cell	Transport system
Red blood cell	Skeletal system
Nerve cell	Reproductive system
Sperm cell	Nervous system

[4]

3. Copy and complete this table by putting a ✔ and a ✗ next to each part of the reproductive system.

Part	Male	Female
Testis	✓	
Egg cell		✓
Vagina		✓
Sperm	✓	
Penis	✓	

[5]

4. Write down the differences between insect and wind pollinated flowers. [3]

5. Insects are important for our food supply.

   Explain why. [2]

6. Pollination requires the transfer of pollen from one flower to another.
   This means the grains have to be very small.

   Describe how you would use a light microscope to look at pollen grains. [6]

Eating, Drinking and Breathing

1. Look at the diagram. It shows how we breathe in and out. Use the diagram to explain what is happening when we:
   - breathe in [3]
   - breathe out [3]

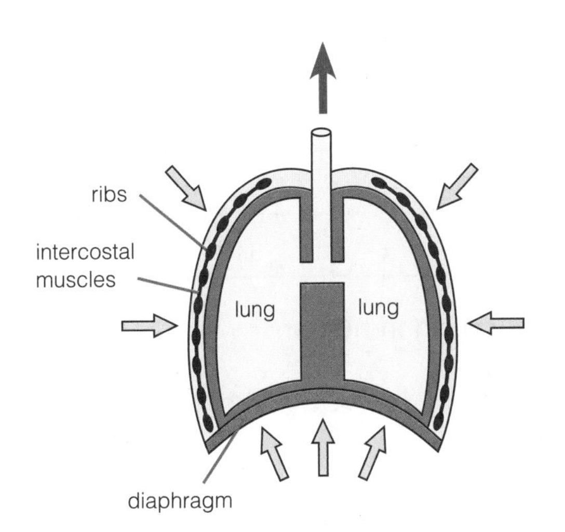

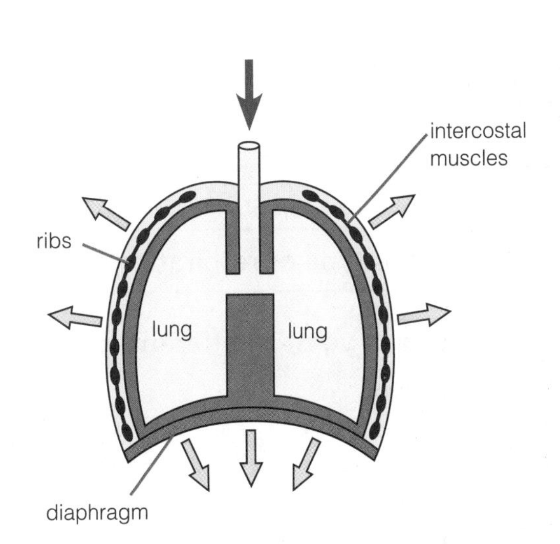

2. Write down three different things that can affect the human breathing system. [3]

Exercise, smoking, asthma

3. Write down why a balanced diet should contain each of the following things:
   carbohydrate, fat, protein, vitamins, minerals, fibre, water [7]

4. Explain the importance of bacteria in the digestive system. [2]

5. An enzyme is a biological catalyst. Explain what this means. [2]

6. The following statements describe the roles of some organs associated with the digestive system.
   One of the statements is incorrect.
   Copy the table and put a cross (✗) against the incorrect statement.

The oesophagus joins the mouth to the stomach	
The stomach produces an acid to break down food	
The small intestine is where water is absorbed	
Waste material leaves the body through the anus	

[1]

Getting the Energy your Body Needs

1. Copy and complete the table by putting a tick (✔) in the correct box next to each statement about respiration.

	Aerobic	Anaerobic
Uses oxygen	✓	
Produces lactic acid		
Produces alcohol		✓
Releases the most energy		
Fermentation uses this type of respiration		✓

[5]

2. Anaerobic respiration in yeast is different to anaerobic respiration in humans. Describe the differences. [3]

3. Explain the importance of having a skeleton. [4]

4. Look at the diagram of the skeleton.

Complete the labels. [10]

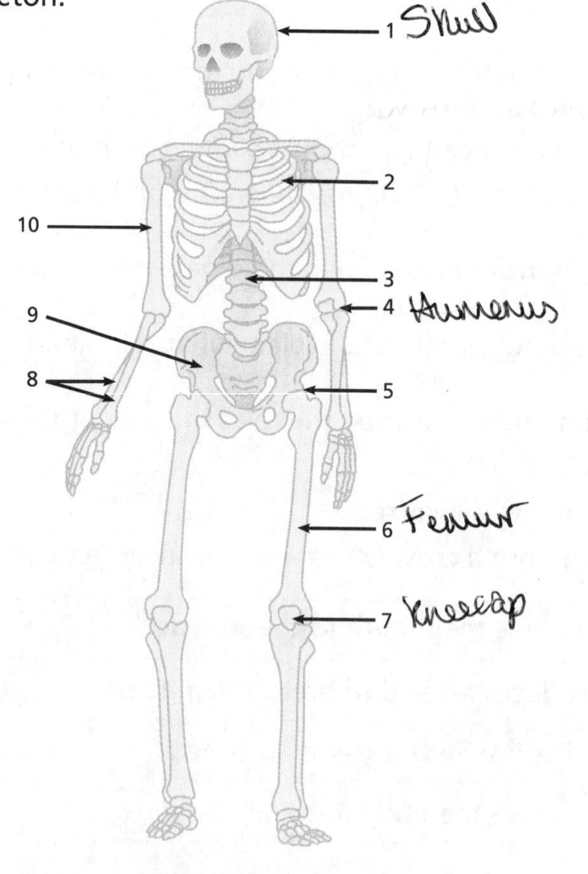

5. Explain why each joint needs at least two different muscles. [3]

Looking at Plants and Ecosystems

1. Copy and complete the table by putting a tick (✔) in the correct box next to each statement about respiration and photosynthesis.

	Respiration	Photosynthesis
Produces oxygen		
Produces carbon dioxide		
Uses energy from sunlight		
Releases energy		
Requires chlorophyll		

[5]

2. Explain how a green leaf is adapted to perform photosynthesis. [5]

3. Complete this food web by adding arrows to show the flow of energy through the web. [7]

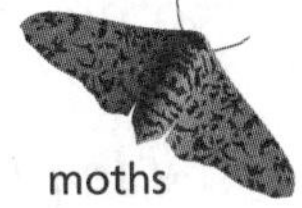

4. Explain why insects are so important in the production of food for humans. [2]

5. Poisonous substances sometimes get released into the environment. These poisons can be more dangerous to animals at the top of the food chain. Explain why. [2]

Variation for Survival

You must be able to:

- Explain how genetic information is inherited from our parents
- Understand the job of chromosomes and genes
- Know the contribution of different scientists to our understanding of DNA.

How Genetic Information is Inherited

- We inherit half our genetic information from our mother and half from our father. This is called **heredity**.
- The **inheritance** of genetic information happens when a sperm from the father fertilises an egg from the mother.

Fertilisation of egg

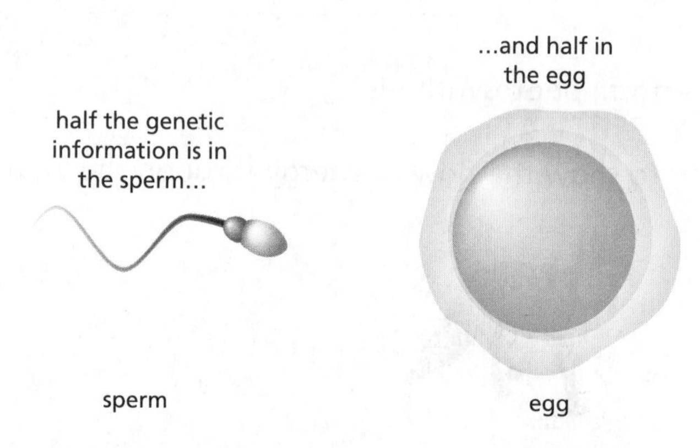

Key Point

Half our genes come from our father and half from our mother. This is why we have similarities to both our mum and dad.

- The genetic information is stored on **chromosomes**, found in the nucleus of our cells.
- The nucleus of almost every cell in our body contains 46 chromosomes.
- However, sperm and eggs only contain 23 chromosomes.
- So 23 chromosomes come from our mother and 23 from our father.
- This produces new offspring with some features inherited from our mother and some from our father.

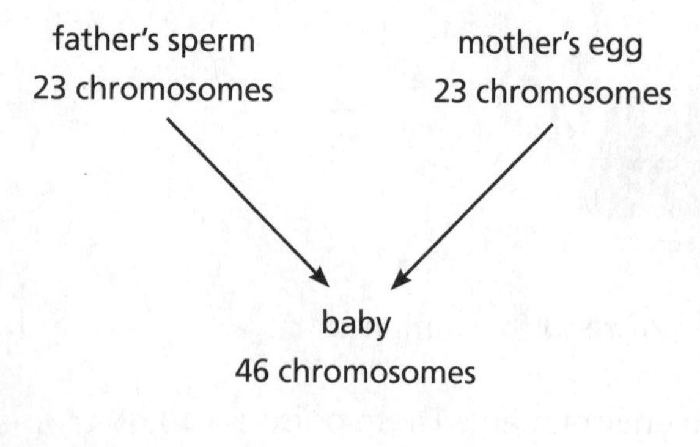

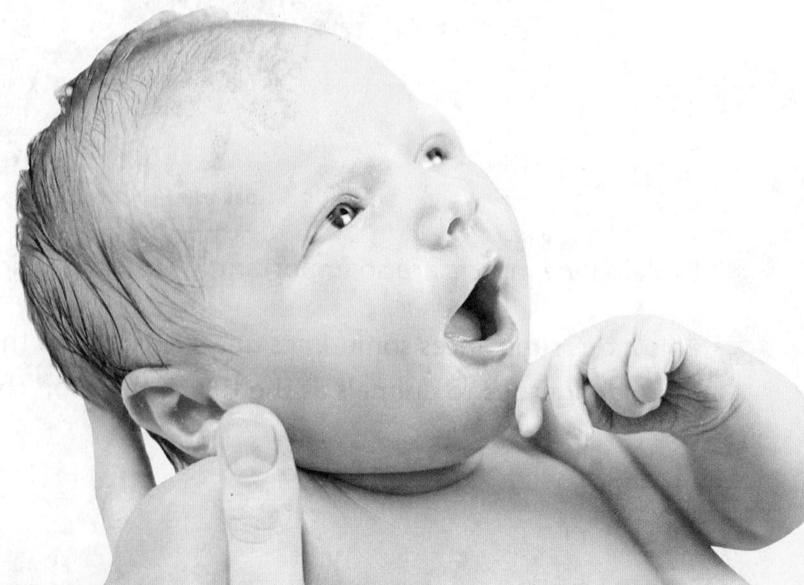

- Each chromosome consists of a very long strand of **DNA**.
- The DNA is divided up into single units called **genes**.
- Each gene is a single section of DNA that codes for a protein.

From cell to gene

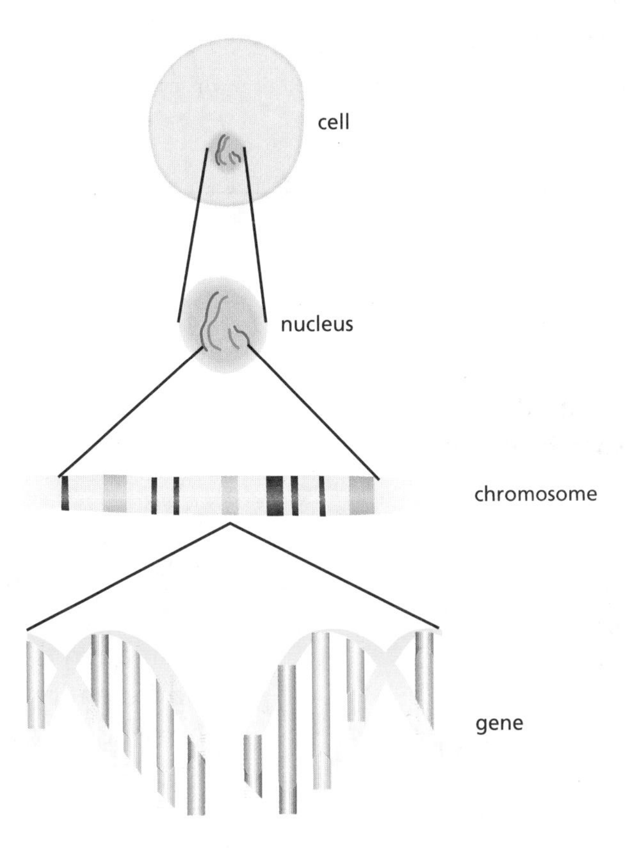

Key Point

A gene is an instruction. It tells the cell how to make a specific protein needed by the cell.

A gene

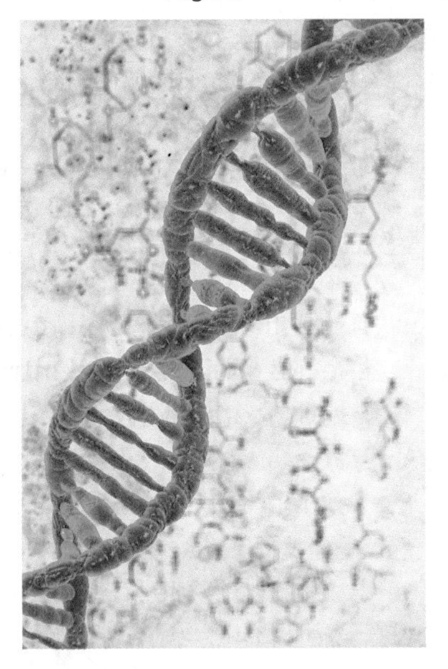

Famous Scientists and DNA

- Several different scientists played a part in discovering how genetic information is inherited.
- In 1953, James Watson and Francis Crick developed a theory for the structure of DNA.
- Maurice Wilkins helped produce evidence to support Watson and Crick's theory of the structure of DNA.
- Rosalind Franklin made X-ray images of DNA which showed that DNA was a double helix.
- Watson, Crick and Wilkins were awarded the Nobel prize for their work. Sadly, Franklin had died several years earlier and so could not be awarded the prize, even though her work was crucial to the discovery of DNA structure.

Key Point

Most scientific discoveries are a result of scientists working together and building on ideas from other scientists.

Quick Test

1. Where does a baby get its 46 chromosomes from?
2. Which is larger, a chromosome or a gene?
3. Write down what a gene is.
4. Name four scientists who were involved in the discovery of DNA.

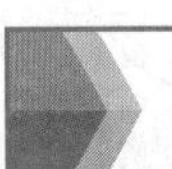

Key Words

heredity
inheritance
chromosome
DNA
gene

Biology

Variation for Survival

You must be able to:
- Explain why variation within a species is so important
- Understand different types of variation
- Explain the effect of a changing environment on our survival.

Variation Between Species

- A **species** is a group of organisms that can reproduce to produce **fertile** offspring.
- All species are different. Scientists call this **variation**.
- Variation between species is called **interspecific variation**.

Variation Within a Species

- As well as variation between species, variation occurs *within* a species. This is called **intraspecific variation**.
- Apart from twins, all humans look different.
- Variation occurs because of sexual reproduction.
- Sexual reproduction mixes up the genes from mum and dad and this causes variation.
- Apart from identical twins, no two brothers or sisters will inherit the same combination of genes from their parents.

Key Point

Variation is due to different organisms having a different combination of genes.

Types of Variation

- There are two types of variation within a species: **continuous variation and discontinuous variation.**
- Height is an example of continuous variation. Some people are tall, others are short. But most people are somewhere in between.

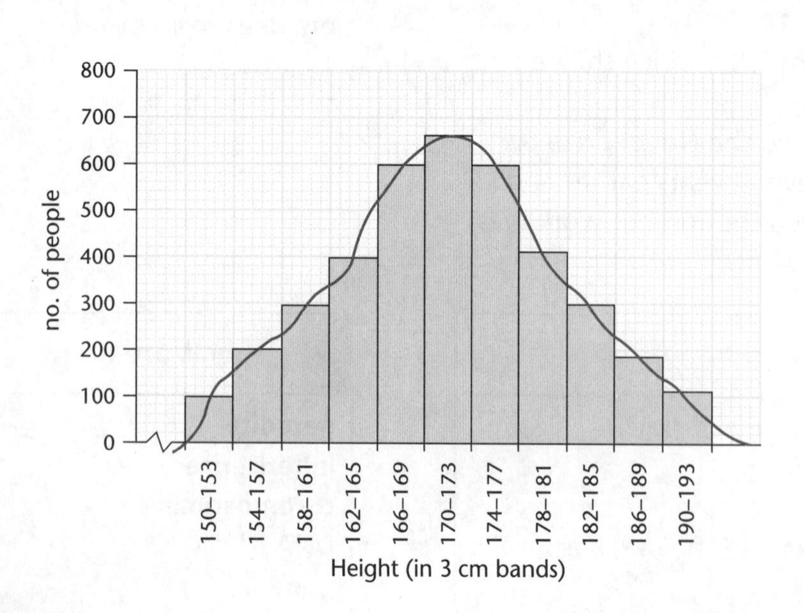

- Blood groups are an example of discontinuous variation. People are either A, B, AB or O. There is no gradual range that goes from one to another.

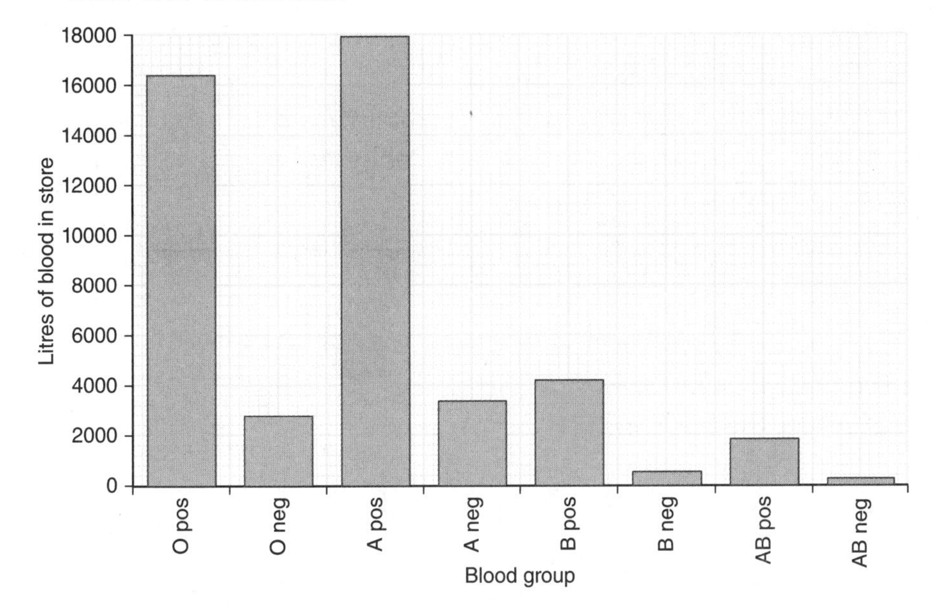

The Effect of a Changing Environment

- Variation is very important when the environment changes.
- **Extinction** of a species can happen when the environment changes. For example, global warming may cause some species to become extinct.
- This is because they are less able to compete for resources and reproduce in the changing environment.
- However, if the changes to the environment are small or occur slowly, because the members of a species are all slightly different some will be able to survive.
- These individuals survive and breed to produce new offspring, ensuring the survival of the species.
- This is why biodiversity is so important.
- The more biodiversity, the less likely it is that extinction will happen.

Key Point

Biodiversity is a measure of the amount of variation between different organisms.

Gene Banks

- A gene bank is a place where scientists store seeds and cells from as many different organisms as possible.
- This helps to ensure that no genes are lost during extinction.
- These genes may be helpful in the future to provide new medicines or food.

Key Words

species
fertile
variation
interspecific variation
intraspecific variation
continuous variation
discontinuous variation
extinction
biodiversity

Quick Test

1. Write down and explain two different types of variation.
2. Write down the main cause of variation in humans.
3. Explain the meaning of the word biodiversity.
4. What is a gene bank?

Our Health and the Effects of Drugs

You must be able to:

- Recall the four main types of drugs
- Describe legal and illegal recreational drugs and their effects
- Describe the dangers of smoking and drinking alcohol
- Understand drug addiction and withdrawal.

Main Types of Drugs

Key Point

Drugs can be used for medical or recreational reasons.

- A **drug** is a substance that affects the body in some way.
- There are many different drugs that affect the body in different ways.
- The four main types of drug are painkillers, depressants, stimulants and hallucinogens:

Type of drug	Effects	Examples	Dangers
Painkillers	Reduce pain and inflammation	Aspirin Paracetamol	Excess may damage stomach lining
Depressants	Make a person feel relaxed and drowsy	Cannabis Heroin	Drowsiness and lack of coordination; can cause post-use hallucinations Reduces breathing and can cause death
Stimulants	Make a person feel energetic and alert	Cocaine Amphetamines	Can cause aggression and paranoia Can cause depression and panic
Hallucinogens	Make a person hear and see things more intensely	LSD Magic mushrooms	Confusion; users see and hear things that are not there Can cause a bad trip; can cause flashbacks for some time afterwards

- All of these types of drug can be addictive.
- All drugs have **side-effects**.
- Side-effects are unwanted symptoms caused by taking the drug, e.g. rashes, headaches, or nausea.

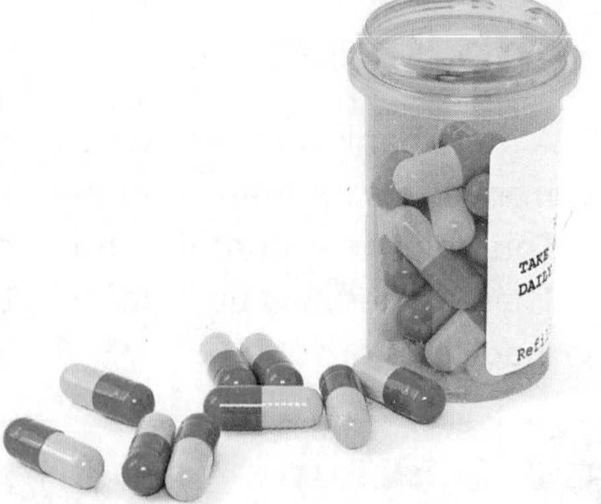

Recreational Drugs

- Drugs that are taken for non-medical reasons are called **recreational drugs**.
- Some recreational drugs are legal and are in common use, for example, caffeine, tobacco, and alcohol.
- Some recreational drugs are illegal and can have dangerous side-effects, for example, cannabis, ecstasy and cocaine.

Smoking and Drinking Alcohol

- Two common recreational drugs are:
  - alcohol (found in alcoholic drinks such as wine and beer)
  - tobacco (found in cigarettes).
- People smoke cigarettes and drink alcohol as a way of relaxing and feeling more confident.
- Smoking and drinking too much alcohol carry serious health risks:

Smoking	Tobacco contains nicotine, a drug which speeds up heart rate and raises blood pressure, leading to increased risk of heart disease, heart attacks and strokes Smoking damages blood vessels and the lungs, leading to coughs, lung infections such as bronchitis, emphysema and lung cancer
Drinking alcohol	In the short term, alcohol: – slows down reactions – reduces coordination – can alter people's behaviour. In the long term, too much alcohol can cause: – liver failure – brain damage – increased risk of strokes and heart attacks – anxiety and depression

Addiction and Withdrawal

- **Addiction** to a drug means that when the person stops taking the drug, they suffer withdrawal symptoms.
- These symptoms may include sweating, shivering, headaches, muscle pain and sickness.

Quick Test

1. Write down a definition for the word 'drug'.
2. Explain what is meant by 'side-effect'.
3. Give two examples of legal recreational drugs.
4. What does the word 'addiction' mean?

Key Words

drug
side-effect
recreational drug
addiction

Biology

Our Health and the Effects of Drugs

You must be able to:
- Explain how microbes can cause disease
- Describe how the body acts as a barrier to prevent disease
- Understand how bacteria, viruses and fungi cause disease
- Understand the importance of vaccination and antibiotics.

Microbes

- Microbes or microorganisms are very small organisms that can only be seen by using a microscope.
- Most microbes are harmless to humans but a small number of them can cause disease.

Key Point

Some microbes are even useful to humans, such as yeast used for bread and wine making.

How Microbes Cause Disease

- Microbes can cause disease in one of two ways:
  1. They can attack and destroy cells in our body.
  2. They can produce chemicals called **toxins** that act like poisons in our body.
- Different types of microbes produce different types of diseases in our body.

Key Point

Microbes that cause disease are called **pathogens**.

How the Body Protects us from Disease

- Fortunately our body can protect us from some diseases. It can do this in several different ways.
- The skin acts as a barrier to stop the microbes entering the body.
- Microbes try to enter the body though body openings.

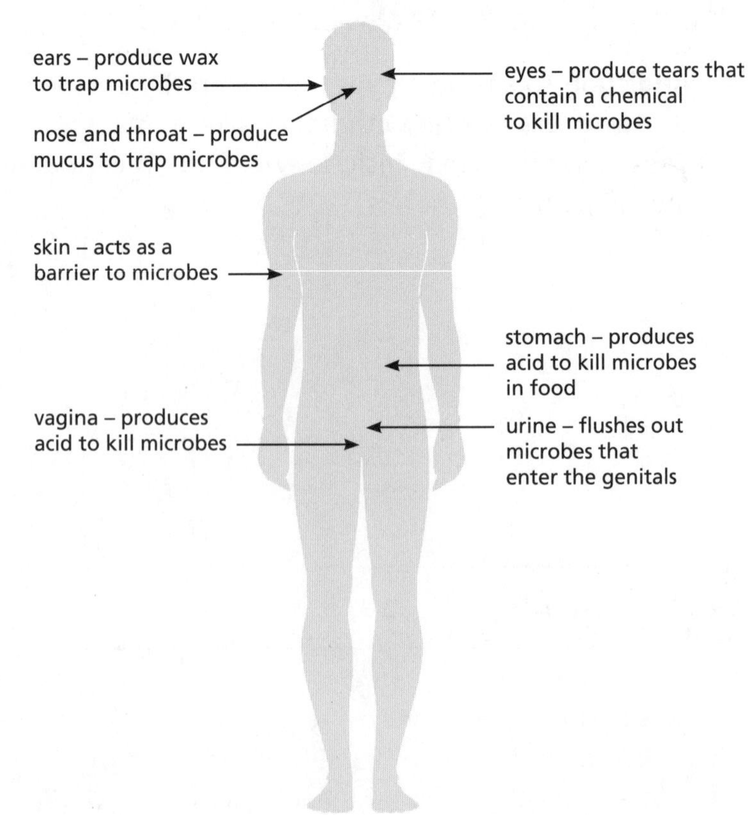

How the Blood Protects us Against Microbes

- Sometimes the skin gets damaged and microbes gain entry. We then need a different kind of defence against microbes.
- Our blood can clot to stop microbes from getting into the blood.
- Our blood contains white blood cells. These cells can attack and engulf microbes.

- They can also make microbes clump together and produce chemicals to destroy them:

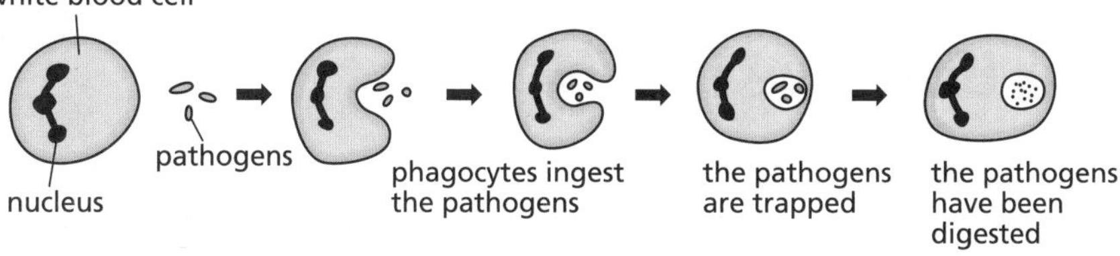

- There are some white blood cells called **memory cells**. When we get a disease that we recover from, our body makes memory cells. If the same microbe enters the body again the memory cells produce **antibodies** to destroy it even before we are aware of the fact that we are ill.

Key Point

White blood cells are part of a defence system called the immune system.

Bacteria, Viruses and Fungi

- There are three different types of microbes that can cause disease, **bacteria**, **viruses** and **fungi**.
- Bacteria can be seen with a light microscope and cause diseases such as tuberculosis.
- Viruses are much smaller and can only be seen with an electron microscope. They cause diseases such as polio.
- Fungi cause diseases such as athlete's foot.

Vaccines and Antibiotics

- Some diseases can be prevented by **vaccination** and some can be cured by **antibiotics**.
- Vaccination is when dead microbes are injected into the body causing the blood to make memory cells. We are then protected against that microbe.
- Antibiotics are chemicals which kill bacteria that have entered our body.

Key Point

Antibiotics do not work against viruses.

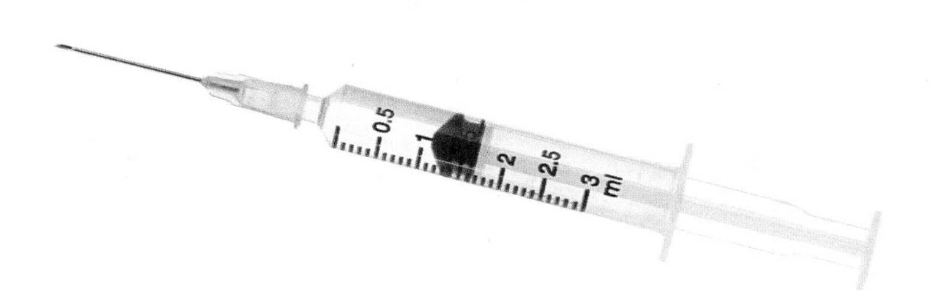

Quick Test

1. Describe three ways our body stops microbes from entering.
2. Name three different types of microbes.
3. Explain what is meant by vaccination.
4. Explain why doctors do not prescribe antibiotics for infections caused by a virus.

Key Words

toxins
pathogen
memory cell
antibodies
bacteria
virus
fungi
vaccination
antibiotic

Getting the Energy your Body Needs

1 Copy the table below and draw a straight line from each description of respiration, to the correct type of respiration.

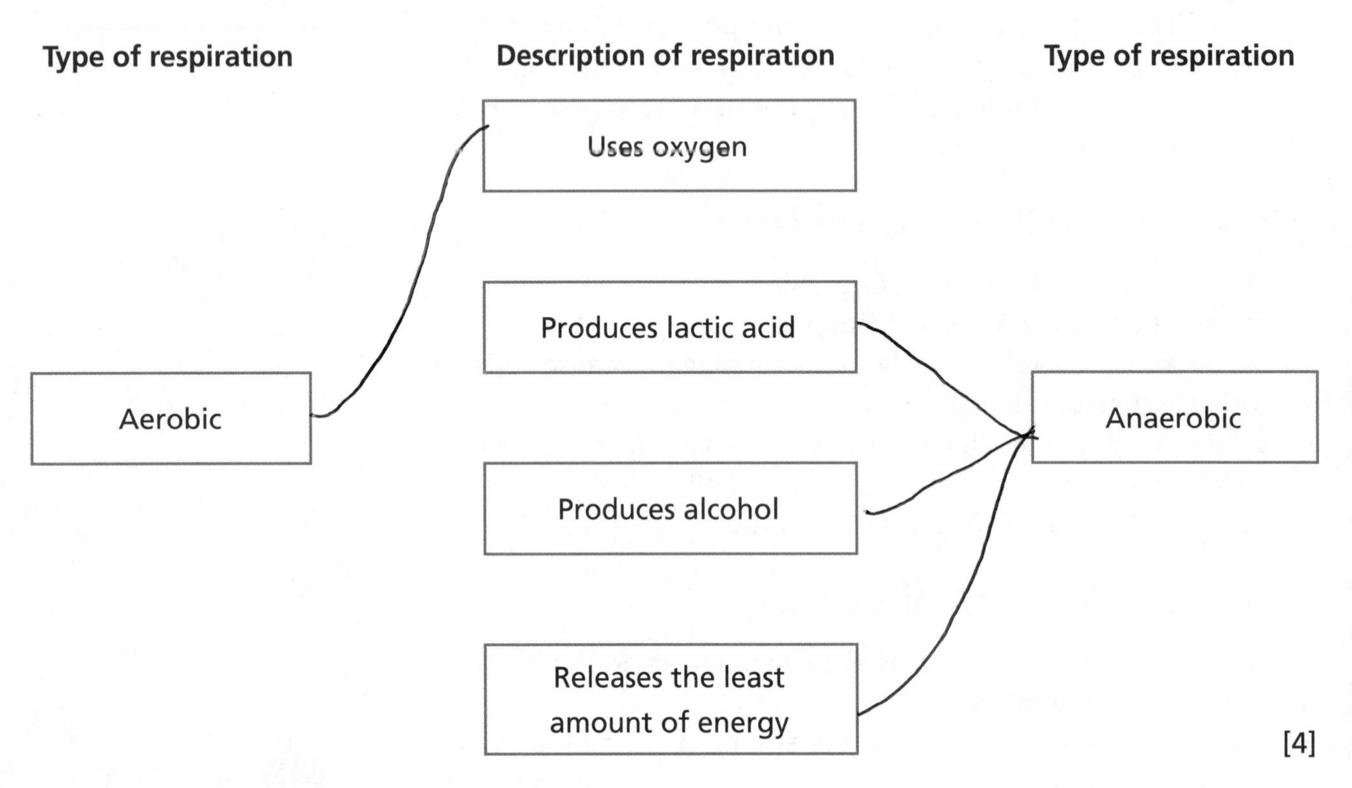

[4]

2 Complete the following word equations:

a) for aerobic respiration

oxygen + glucose .... → water + carbon dioxide .... + energy [2]

b) for fermentation in yeast

glucose → alcohol .... + carbon dioxide .... + energy [2]

c) for anaerobic respiration in humans

glucose → respirations .... + energy [1]

3 Explain why respiration in living organisms is so important. [2]

Answer = 129 page

4 The skeleton is an important structure.

Copy the table and put a tick (✓) in the boxes next to each function performed by the skeleton.

Carries oxygen around the body	
Supports the body	✓
Helps with movement	✓
Where food is digested	
Protects some organs	✓
Makes red blood cells	✓
Where anaerobic respiration takes place	

[4]

5 Joints allow the skeleton to move. Identify the structures numbered 1–5 in the diagram of the knee joint. [5]

1. cartilage

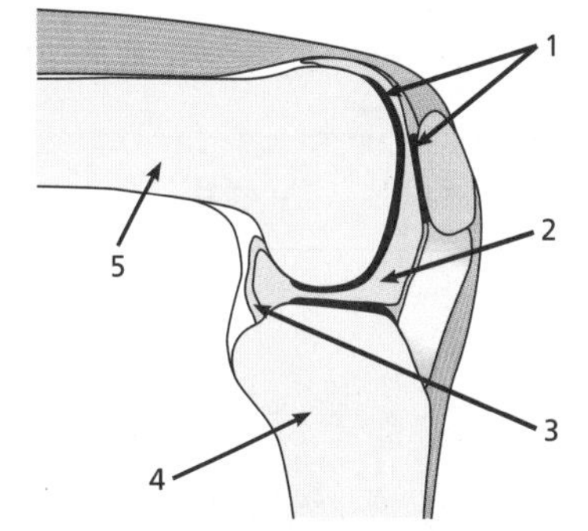

6 Look at the diagram of the human arm.

a) Explain the job done by organ **A**. [2]

b) Explain the job done by organ **B**. [2]

c) What single word best describes these two organs? [1]

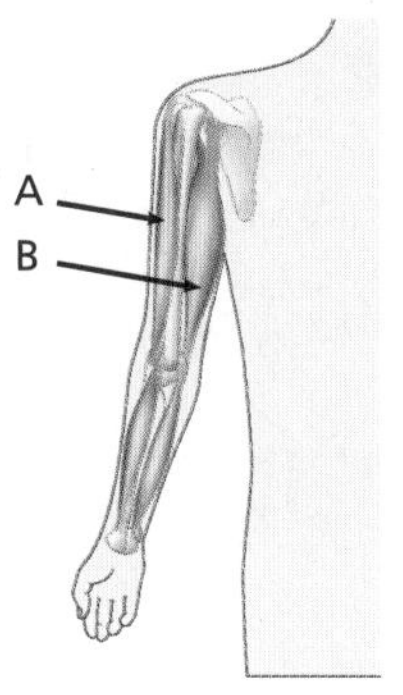

Looking at Plants and Ecosystems

1 **a)** Complete this word equation for photosynthesis:

water + carbon dioxide → glucose + oxygen [2]

b) Write down two other things needed for photosynthesis to take place. [2]

2 Using the two gases, carbon dioxide and oxygen, show how animals and plants are dependent upon each other. [4]

Variation for Survival

1. Copy and complete the table below by writing down in each box the correct number of chromosomes found in the nucleus of these types of human cell.

Cell	Number of chromosomes
Muscle cell	46
Nerve cell	46
Sperm cell	23
Egg cell	23
Embryo cell	46

[5]

2. Give the correct labels for **A** to **D** on the diagram. Choose from the words given below. [4]

nucleus	cell	gene	chromosome

A cell

B nucleus

C chrosome

D gene

3 The graph shows variation of a characteristic found in humans. Variation can be either continuous or discontinuous.

Use the graph to explain the differences between these two types of variation. [2] 

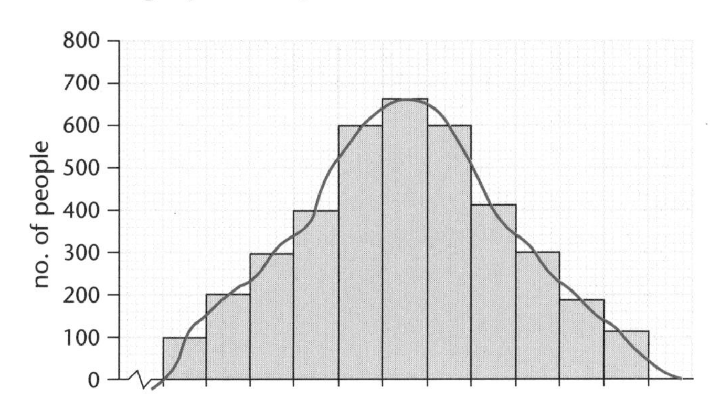

Our Health and the Effects of Drugs

1 Look at the outline drawing of the human body.

Add labels to the drawing to show how the body prevents the entry of microbes. [7] 

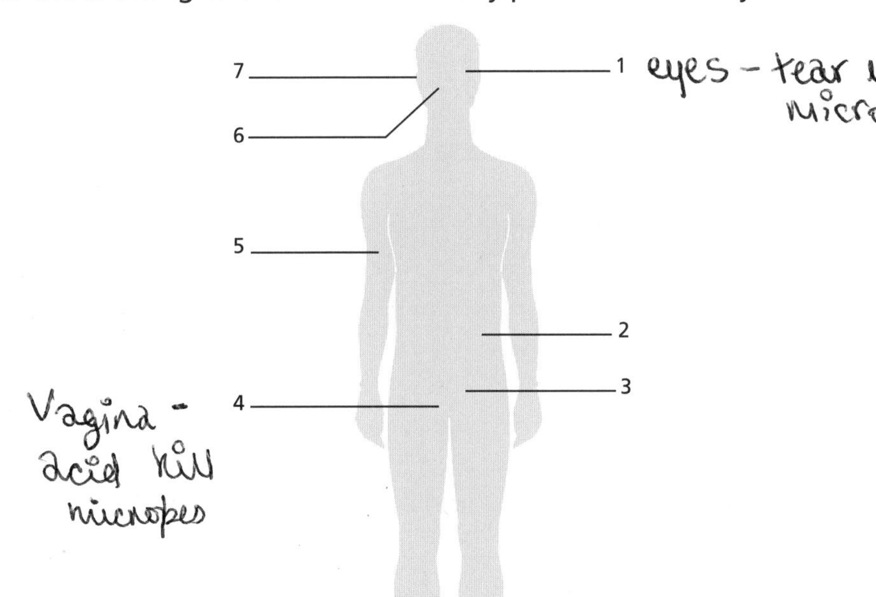

2 Explain the difference between a medical drug and a recreational drug. [2]

3 Recreational drugs can be divided into three categories: depressants, stimulants and hallucinogens. Give one example of each type of drug and describe its effects on the human body. [6]

4 Many people drink alcohol excessively and smoke. Describe the possible long-term effects on the human body of drinking alcohol excessively and smoking. [5]

Chemistry

Mixing, Dissolving and Separating

You must be able to:

- Represent pure substances and mixtures using simple particle pictures
- Demonstrate a range of laboratory skills
- Apply appropriate separation techniques to different mixtures.

Pure and Impure Substances

- In chemistry a 'pure' substance is one that contains only one type of atom or compound.
- An impure substance contains more than one substance (element or compound), forming a mixture.
- The substances in the mixture are not chemically joined together so it should be easy to separate them.
- A common example of a mixture is sugar dissolved in water.

Particle diagram of sugar dissolved in water

Water molecules in a pure solution

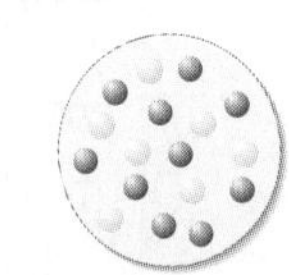

A sugar solution with water and sugar molecules

Chromatography

- Chromatography separates dissolved pigments in solution, e.g. the pigments in ink:
  1. The sample mixture is loaded on a pre-marked line at the bottom of a piece of chromatography paper and dipped into solvent.
  2. As the **solvent** moves up the paper it takes the dissolved pigments with it.
  3. Since the pigments have different solubilities they travel at different speeds and so separate.
  4. The most soluble pigments move the furthest; less soluble pigments move less far.

Chromatography

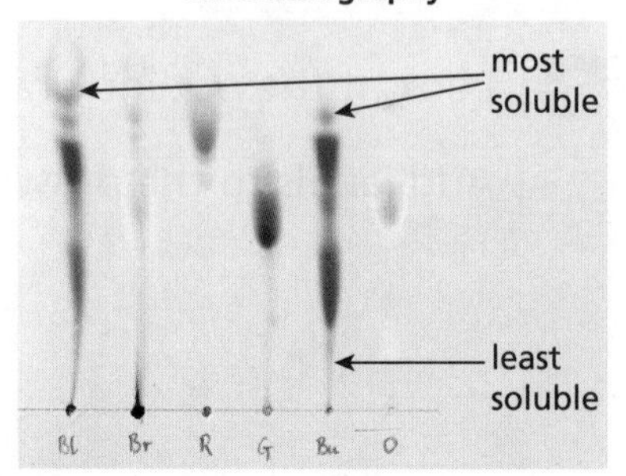

Key Point

A pencil line marks the starting point as it will not move with the ink pigments.

Filtering

- **Filtration** separates an insoluble solid from a liquid by passing the solid/liquid mixture through filter paper.
- The **filtrate** is the liquid which passes through the filter paper and the solid left behind is the residue.
- Excess copper oxide in copper sulfate solution can be separated by filtration. Copper sulfate is the filtrate, copper oxide is the residue.

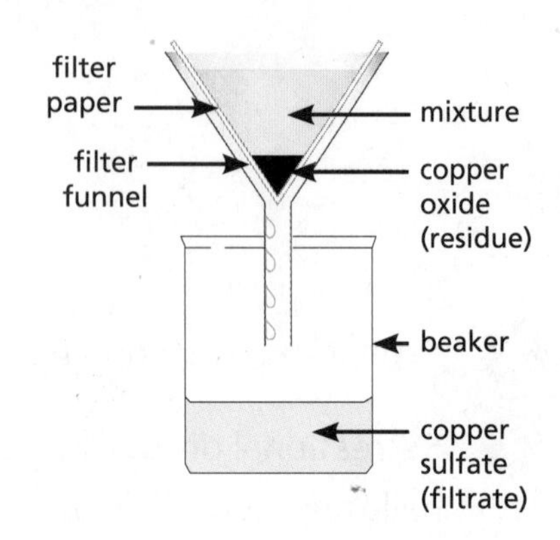

Evaporation

- **Evaporation** is used to remove the liquid part of a mixture and collect the dissolved solid.
- The mixture is placed in a suitable container (e.g. a watch glass) and heated, sometimes by using a Bunsen burner.

Key Point

The slower the liquid evaporates, the larger the crystals that form.

Distillation

- Liquids have different boiling points.
- By carefully controlling the temperature of a heated mixture of two or more liquids, the liquids evaporate at different times. This is known as distillation.
- The evaporated gas is cooled back into liquid and collected as **distillate** in a collecting vessel.
- The fragrances used in perfumes are separated by distillation, as well as the different parts of crude oil.

Distillation of ethanol (an alcohol) and water mixture

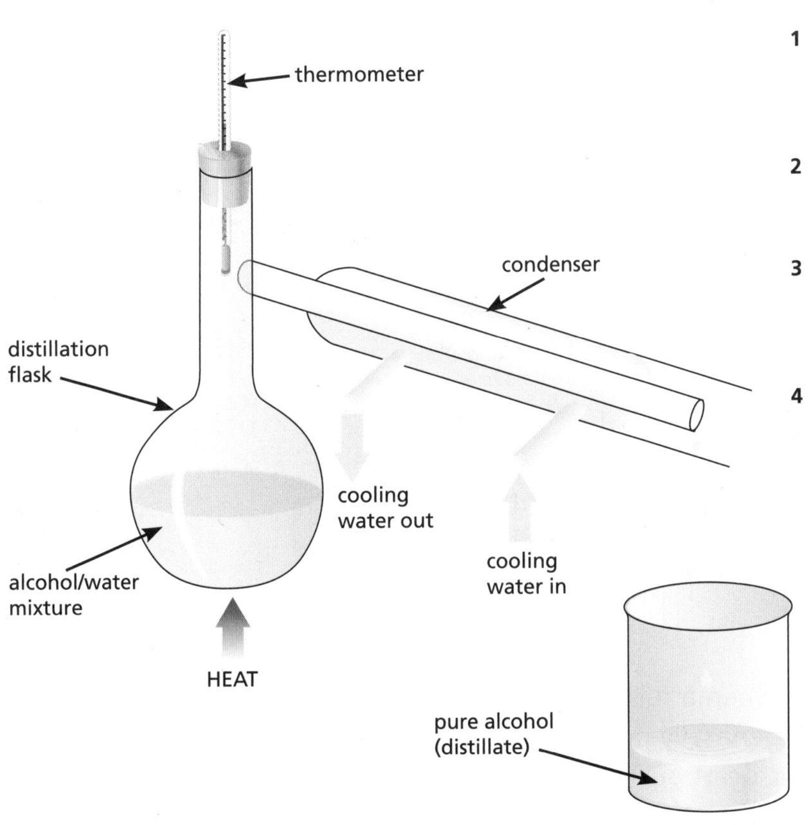

1 The mixture is heated until the liquid with the lowest boiling point boils.

2 The thermometer measures the temperature of the gas.

3 The water in the condenser cools the gas, allowing it to condense back into a liquid.

4 The liquid (distillate) is collected.

Quick Test

1. How could you separate an insoluble solid from a liquid?
2. What size crystals are made from rapid evaporation?
3. What does 'distillate' mean?
4. Describe how to carry out chromatography.

Key Words

solvent
filtration
filtrate
evaporation
distillate

Chemistry

Mixing, Dissolving and Separating

You must be able to:

- Explain the conservation of mass in reactions and changes of state
- Represent pure substances and mixtures using particle pictures and word equations
- Explain similarities and differences between combustion, thermal decomposition, oxidation and reduction.

Conservation of Mass

- The law of conservation of mass states that in any physical change or chemical reaction the mass after the change will be the same as the mass before the change.
- With state changes this means that the number of particles of the substance at the start will equal the number of particles at the end.

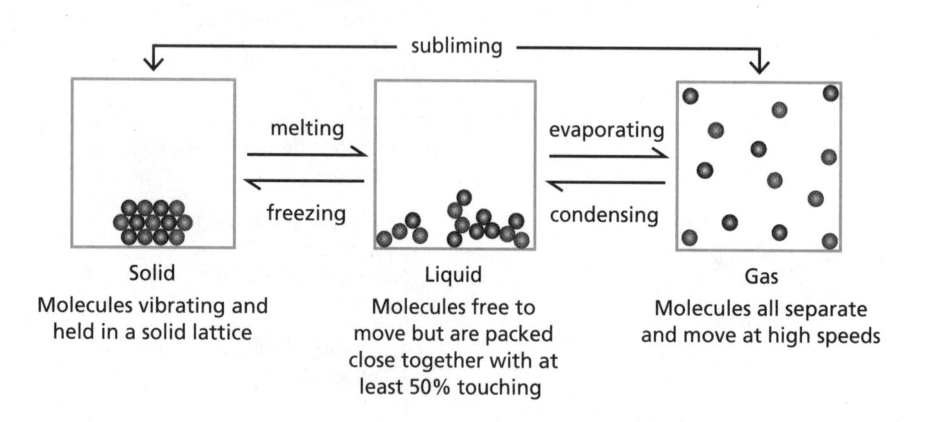

Key Point

When drawing a particle diagram for a liquid at least half of the particles should be touching each other.

- With a chemical reaction the atoms of the reactants are rearranged to form the products. Atoms cannot 'disappear'.

Combustion

- **Combustion** is the reaction between a fuel and oxygen.
- Carbon dioxide and water are generally produced as waste products when the fuel is a hydrocarbon.
- Energy is released as heat and light.

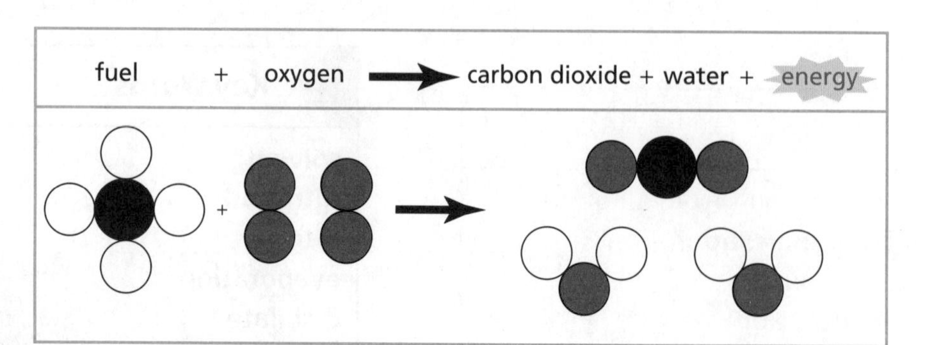

Key Point

Energy has not been made, it has just been released.

Thermal Decomposition

- Some compounds break down into new molecules when heated; they don't react with oxygen in the air.
- This is called **thermal decomposition**.
- An example is chalk, which has the chemical name calcium carbonate.

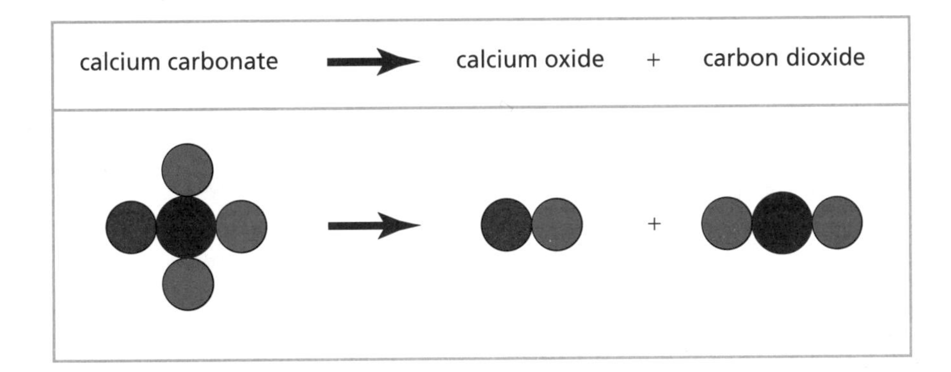

Key Point

The coloured circles represent atoms. There are the same number on each side of the equation.

Oxidation and Reduction

- When substances gain oxygen in a reaction it is called **oxidation**.
- Losing oxygen in a reaction is called **reduction**.
- For example, carbon can be oxidized to form carbon dioxide:

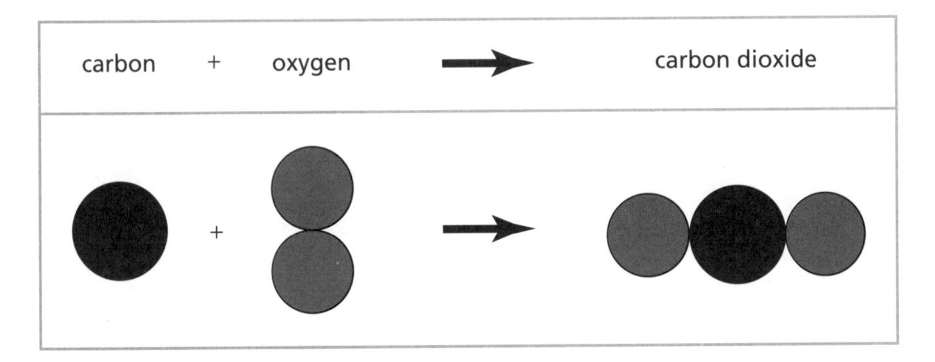

- The reaction of iron with water and oxygen is a special form of oxidation, forming iron (III) oxide, which is known as **rust**.

iron + water + oxygen → hydrated iron (III) oxide

- Rusting requires oxygen and water. It happens faster when salt is dissolved in the water.

Oxidation forming rust

Key Point

Only use the term rust for the oxidation of iron. Other metals corrode, they don't rust.

Key Words

combustion
thermal decomposition
oxidation
reduction
rust

Quick Test

1. Draw diagrams to show the atoms in a solid, liquid and gas.
2. Describe how calcium carbonate thermally decomposes.
3. What is meant by the term oxidation?
4. What does iron have to react with in order to rust?

Elements, Compounds and Reactions

You must be able to:

- Explain the structure of the periodic table, including groups, periods, symbols and formulae
- Explain differences between elements and compounds in terms of particles
- Describe the differences between physical changes and chemical reactions.

The Periodic Table

- The periodic table contains all the **elements** that are found in the universe.
- An element is a substance that contains only identical atoms.
- The simplest particle of an element that cannot be broken down further without losing its properties is called an **atom**.
- The periodic table arranges the elements based on the atomic number (the number of protons in the nucleus of each atom) and the physical and chemical properties of each element.
- Each column of the periodic table is called a **group**, a family of elements with similar physical and chemical properties.
- The rows in the periodic table are called **periods**. The atomic number increases from left to right through the period.
- The majority of elements in the periodic table are metals; the non-metals are less common.

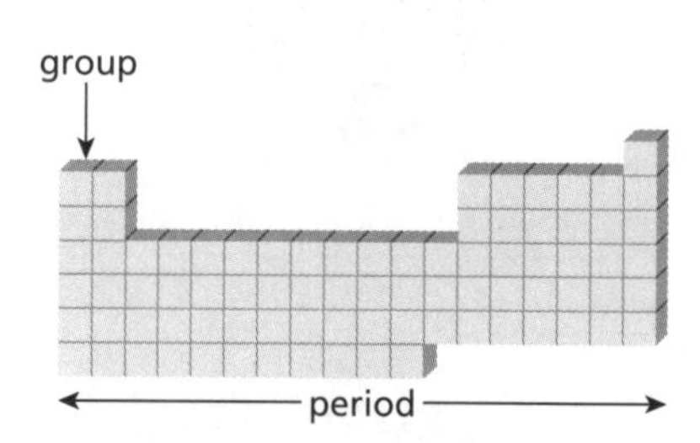

Key Point

The atomic number of elements increases sequentially in whole numbers as you go through the table.

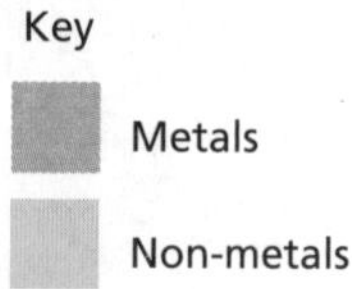

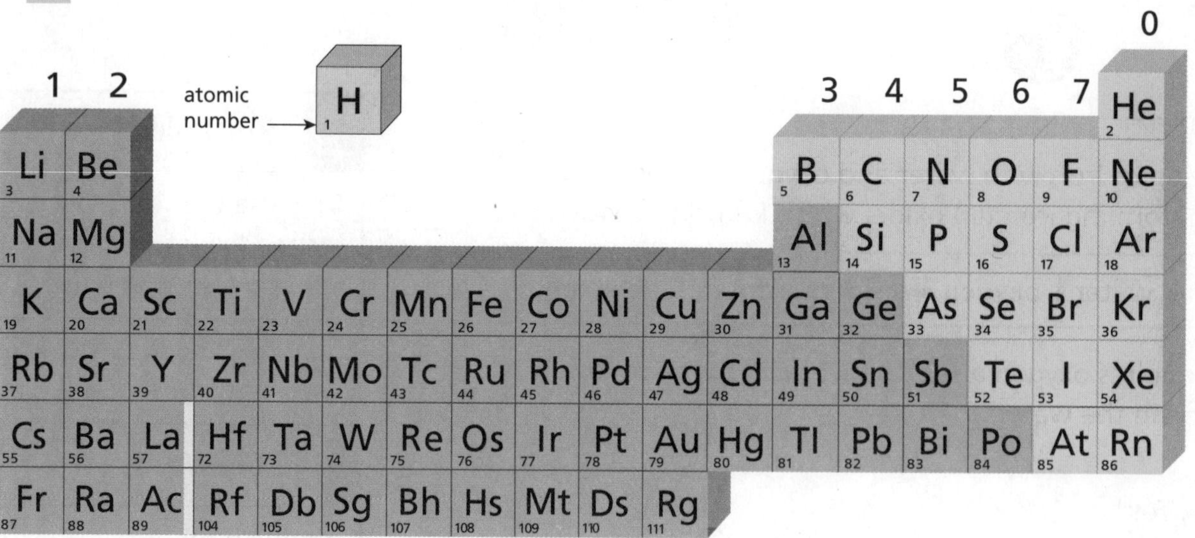

- The periodic table used today was devised by the Russian chemist Dmitri Mendeleev.
- We can use the periodic table to predict the physical and chemical properties of elements.

- For example, elements in a group are very similar. They have similar physical properties and chemical reactions:
  - The metals in group 1 all react with water to form alkaline solutions
  - The non-metals in group 7 are good at killing bacteria.

Chemical Symbols and Formulae

- Elements have a name and a chemical symbol.
- Normally this is one or two letters, for example Helium = He, Copper = Cu, Silver = Ag.
- When chemicals react and chemically join together they form compounds.
- The compound is represented by a chemical formula, e.g. water = H_2O, where two hydrogen atoms are joined to one oxygen atom.
- The number written as a subscript indicates how many of those atoms are in the compound, e.g. $C_6H_{12}O_6$ (glucose) contains 6 carbon, 12 hydrogen and 6 oxygen atoms.

Key Point

Take care to write only the first letter of an element's symbol as a capital letter.

Physical Changes and Chemical Reactions

- A physical change is where a substance changes state, e.g. water (liquid) freezing into ice (solid).
- Physical changes are easy to reverse.
- A chemical reaction is where elements chemically join together to form a compound.
- It is difficult to reverse a chemical reaction.
- The compound formed has different properties to those of each of the original elements.
- To make it clear what is happening in a reaction we write a chemical equation:
  - On the left hand side we write the **reactants**
  - On the right hand side we write the **products** formed.
- An example is the reaction of sodium with chlorine:

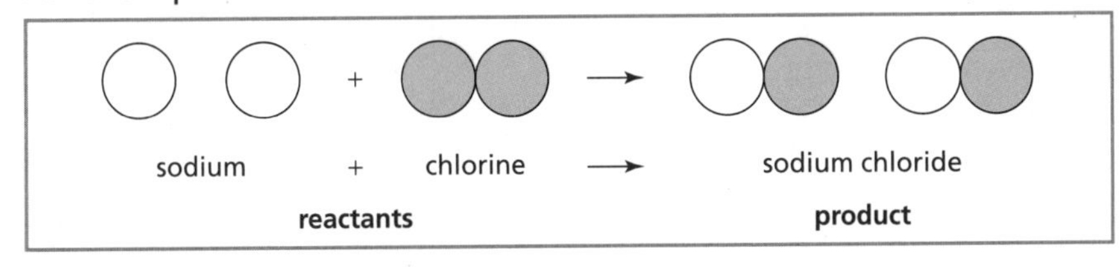

Quick Test

1. Which of the following are compounds?
   O_2 CO_2 H_2O
2. What is the atomic number of an element?
3. How many hydrogen atoms are in H_2SO_4?
4. How many different elements make up $C_6H_{12}O_6$?

Key Words

element
atom
group
period
reactant
product

Chemistry

Elements, Compounds and Reactions

You must be able to:

- Describe the properties of metals and non-metals
- Explain the reactions of metals and metal oxides with acids
- Understand the concept of a reaction using oxidation of metals and non-metals.

Properties of Metals and Non-metals

Metals	Non-metals
Conduct electricity and heat	Are unable to conduct electricity and heat
Are **ductile** (can be drawn into wires)	Often have a low melting point and boiling point
Are **malleable** (can be hammered into shape)	Are often gases at room temperature
Are shiny	Often have a lower **density** than metals.
Are sonorous (ring like a bell when hit)	
Often have a high melting point and boiling point.	

Metal – copper wires used to conduct electricity

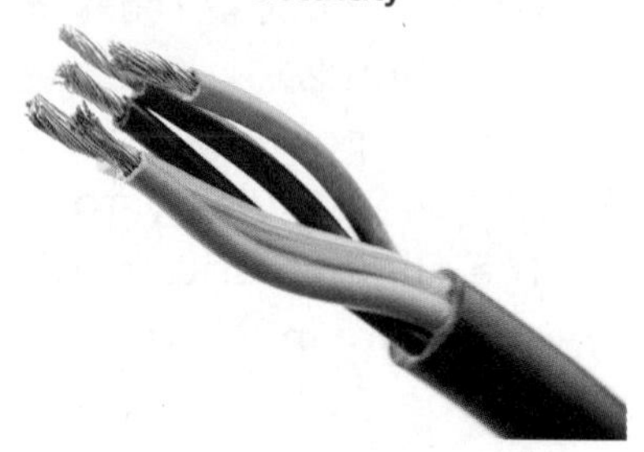

Non-metal – helium gas used to inflate balloons

Reactions of Metals

- Metals react with acids to give a **salt** and hydrogen:

metal + acid ⟶ salt + hydrogen

- The salt formed always takes the name of the metal plus a suffix that represents the acid used in the reaction:

Acid	Suffix	Example
Hydrochloric acid	Chloride	**sodium + hydrochloric acid ⟶ sodium chloride + hydrogen**
Sulfuric acid	Sulfate	**sodium + sulfuric acid ⟶ sodium sulfate + hydrogen**
Nitric acid	Nitrate	**sodium + nitric acid ⟶ sodium nitrate + hydrogen**
Phosphoric acid	Phosphate	**sodium + phosphoric acid ⟶ sodium phosphate + hydrogen**

Key Point

Hydrogen is a gas, so bubbles are always produced when acid and metal react.

Oxidation

- Reacting an element or compound with oxygen is called **oxidation**.
- The atoms that make up the elements and compounds rearrange to make a new compound, an oxide. For example:

magnesium + oxygen → magnesium oxide

- This can also be written as a balanced equation.
- A balanced equation indicates the number of atoms and how they are arranged.
- For example, two atoms of magnesium react with one molecule of oxygen to form two molecules of magnesium oxide:

$$2Mg(s) + O_2(g) \rightarrow 2MgO(s)$$

- Combustion is where a fuel reacts with oxygen (burns) forming carbon dioxide and water, and giving out energy in the process:

fuel + oxygen → carbon dioxide + water + energy

Key Point

The numbers of each atom are always the same on both sides of the formula.

Reactions of Metal Oxides

- Metals react with oxygen to form metal oxides, for example:

magnesium + oxygen → magnesium oxide

- The metal oxide is called a **base** and is the chemical opposite of an acid.
- Metal oxides react with acids to form a salt and water:

metal oxide + acid → salt + water

- This means that the acid has been **neutralised**.
- The salt formed always takes the name of the metal and the suffix from the acid, for example:

sodium oxide + hydrochloric acid → sodium chloride + water

Salt

Quick Test

1. Write the word equation for the reaction between nitrogen and oxygen.
2. What salt is formed in the reaction between potassium and sulfuric acid?
3. Give three properties of a metal.
4. Give three properties of a non-metal.

Key Words

ductile
malleable
density
salt
oxidation
base
neutralise

Variation for Survival

1. The graph shows variation of a characteristic found in humans.

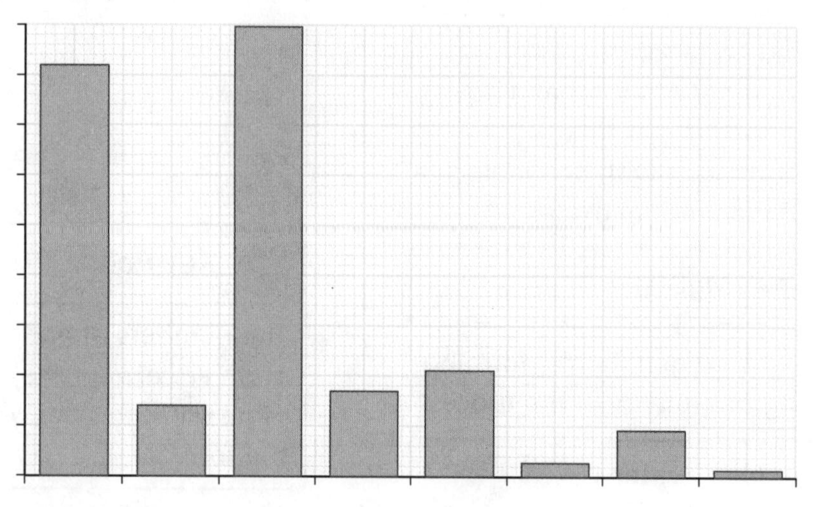

a) What is the name of this type of variation? [1]

b) Write down two examples of this type of variation found in humans. [2]

2. Variation is very important to the survival of a species.

Under which of the following conditions is variation most important? Put a tick next to the best answer from the table below.

When environmental conditions stay the same	
When the environment is changing very slowly	
When the environment is changing very quickly	
The environment has no effect on variation	

[1]

3. Explain the part played by each of the following scientists in the understanding of DNA.

a) Watson and Crick. [1]

b) Rosalind Franklin. [1]

Our Health and the Effects of Drugs

1. Explain the meaning of the following words:

   a) pathogen

   b) toxin

   c) antibody [3]

2. Name three different types of microbe and give an example of a disease caused by each of them. [3]

3. Explain why doctors are less worried about people taking the drug caffeine and more worried about people taking the drug cocaine. [2]

4. Explain the difference between addiction and withdrawal. [2]

5. Match each activity below with a danger of doing it.

Activity	Danger
Drinking alcohol	Trying to fly off a tall building
Smoking	Reduces breathing
Taking LSD	Lung cancer
Using heroin	Liver failure

[4]

6. Microbes sometimes gain entry to our body.

   Explain how white blood cells can protect our body from invasion by microbes. [3]

Mixing, Dissolving and Separating

1. Which of the following separation methods would be best for each of the following investigations?

   Choose from the following methods:

   chromatography **distillation** **filtration**

   a) Extracting alcohol from beer. [1]

   b) Checking whether a note written in blue ink was written using a particular pen. [1]

   c) Separating sand from a mixture of sand and water. [1]

2. **a)** Which two of the following diagrams show a pure substance? [2]

   A

   B

   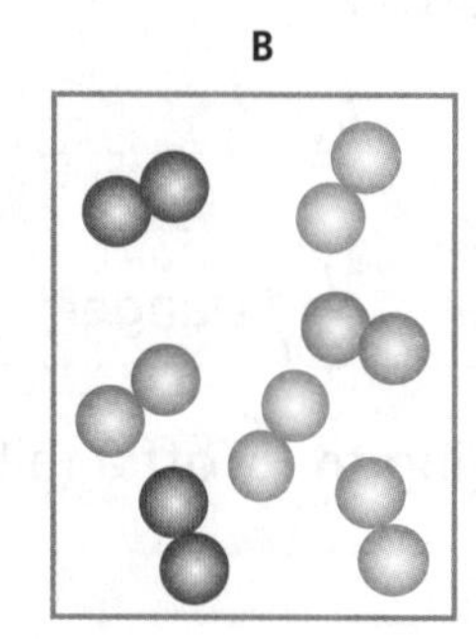

   C

   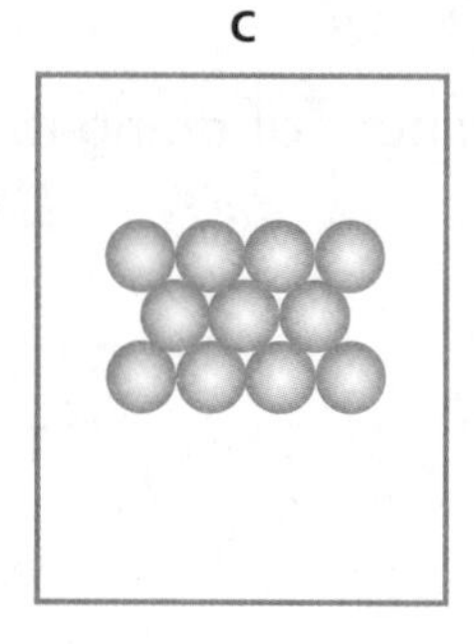

   D

   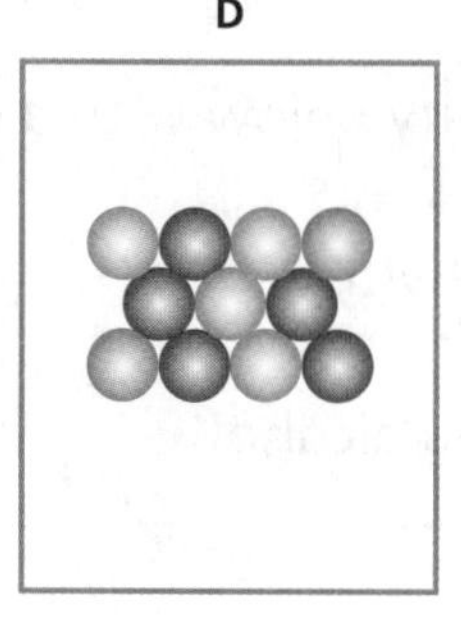

   b) Which of the substances could be copper? [1]

3. Describe how filtration and evaporation could be used to extract salt from seawater. [5]

4. Vinnie is analysing the pigments used to colour different sweets using chromatography.

   Vinnie makes a qualitative observation on how similar the chromatograms are to each other using his eyes and judgement.

   Suggest what Vinnie would need to do to make a more accurate quantitative measurement. [2]

5. An aquarium for keeping fish uses a filter.

   Suggest what the filter is removing from the water and explain how filtration works. [3]

Elements, Compounds and Reactions

1. Ethan is painting a model. When he opens a tin of paint he notices that the paint has separated into layers.

   a) What type of a substance is the paint?

   i) an element **ii)** a compound **iii)** a mixture [1]

   b) When Ethan reads the label on the tin it says that the paint contains water and titanium oxide. Titanium oxide and water are examples of what type of substance?

   i) an element **ii)** a compound **iii)** a mixture [1]

2. Which of the following are examples of chemical reactions?

   i) chocolate melting on a hot day

   ii) a firework exploding in the sky

   iii) bread being toasted

   iv) water condensing on a cold window pane [2]

3. A periodic table is shown below.

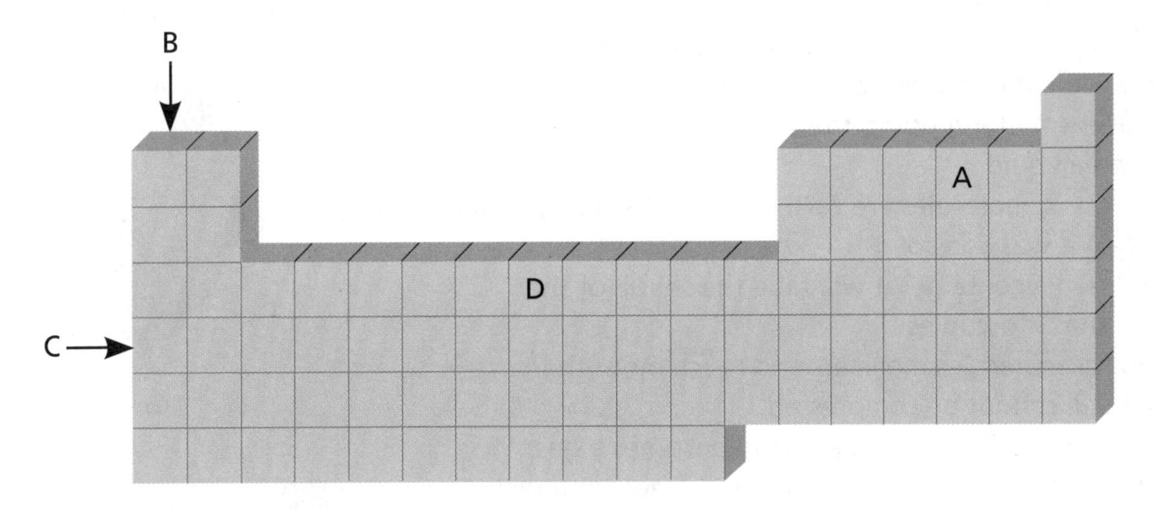

   a) Which letter indicates a group? [1]

   b) Which letter indicates a period? [1]

   c) Which letter indicates a non-metal? [1]

4. Explain the difference between an element and a compound. [3]

Explaining Physical Changes

You must be able to:

- Describe the similarities and differences between solids, liquids and gases
- Explain how changes in temperature affect the motion and spacing of particles
- Explain sublimation in terms of particles.

The Particulate Nature of Matter

- All matter in the universe is made up of **atoms**, arranged in one of three states: solid, liquid or gas.
- At the coldest temperature possible (–273 °C or 0 K), the atoms have no **kinetic energy** so cannot move.
- If heat is introduced, the atoms gain kinetic energy and so move.

Solid

Solids, Liquids and Gases

- Solids:
  - contain atoms arranged as close together as possible
  - are therefore denser than their liquid form (apart from water) and cannot be compressed
  - will have a fixed shape and volume that does not depend upon the container that it is in.
- Even though they form part of a solid the atoms, or molecules, still vibrate due to their kinetic energy.
- As the temperature supplied to a substance increases, the atoms or molecules vibrate more and more.
- Eventually, at the melting point, the atoms or molecules rearrange into a liquid:
  - as the atoms or molecules are further apart the **density** will be less than it was as a solid
  - if a liquid is in a container it will take the shape of the container that it occupies
  - the atoms or molecules move around much more than in a solid, but still cannot be compressed.
- Eventually, at the boiling point, the liquid becomes a gas:
  - the atoms or molecules in a gas can move freely and will occupy all of the available space in a container
  - if not completely enclosed, the gas particles will escape
  - when a material cools, the reverse process happens.
- Unlike solids and liquids, gases can be compressed.
- Kinetic energy is removed from the substance, causing its movement to slow.
- The substance changes from a gas into a liquid (condenses), then into a solid from a liquid (freezing).

Liquid

Gas

Key Point

Gases can still be heated further, to temperatures higher than the boiling point.

Solids	Liquids	Gases
• Particles are touching each other so solids are usually dense. • Particles are kept in place by very strong forces so solids often have a high melting point. • Particles only move by vibrating so solids have a fixed shape.	• Particles are mostly touching each other so liquids cannot be compressed. • Particles are attracted to each other by quite strong forces. • Particles can move over each other so liquids can take the shape of the container.	• Particles are spaced apart so gases can be compressed and have lower densities than liquids and solids. • Particles are attracted to each other by weak forces. • Particles move very quickly.

Sublimation

- Some substances can jump from solid to gas. This is called **sublimation**.
- Examples of sublimation include carbon dioxide (dry ice to gas), ammonium chloride and gel air freshener.

Sublimation

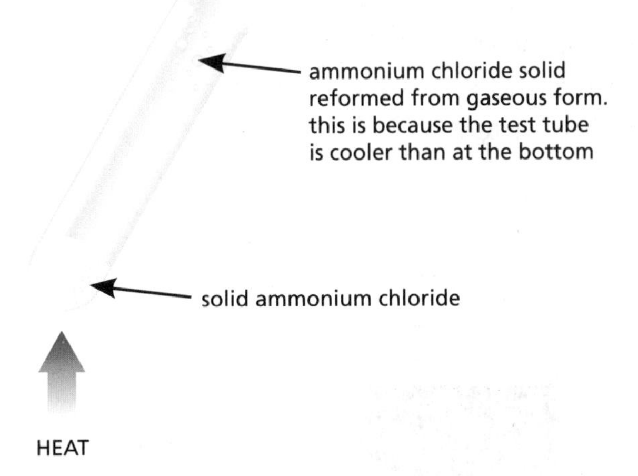

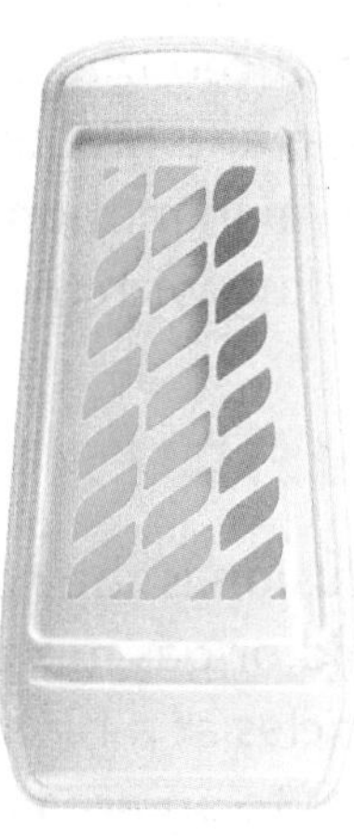

Gel air freshener

Quick Test

1. Why are most solids denser than their liquid form?
2. At what temperature do atoms stop vibrating?
3. Describe what happens in sublimation.
4. State the differences between solids, liquids and gases.

Key Point

Sublimation is the change from a solid direct to a gas, or from a gas to a solid.

Key Words

atom
kinetic energy
density
sublimation

Explaining Physical Changes

You must be able to:

- Describe the particular nature of water and the ice/water transition
- Describe and explain Brownian motion and the diffusion of gases
- Explain the process of heat conduction between particles in a conductor.

Water

- Water has a number of properties that are unique.
- When ice forms, the water molecules line up in a regular pattern.
- The water molecules are further apart in ice than in the liquid form, and therefore solid water is less dense than liquid water.
- Consequently, ice floats on water.

Brownian Motion

- In the 1800s Robert Brown observed pollen grains suspended in water under the microscope.
- He noticed that the particles were moving randomly in the water, and his observation is now called **Brownian motion**.
- Brownian motion is due to the particles suspended in a fluid colliding with the atoms or molecules that make up the fluid.

Key Point

The particle appears to move on its own, this is because the water molecules are so small that they are invisible.

Diffusion

- **Diffusion** is the name of the process whereby molecules in a liquid or gas mix as a result of their random motion.
- Particles at a high **concentration** in one location will tend to move to an area where they are in low concentration.
- Eventually the particles will become evenly distributed throughout the liquid or gas.

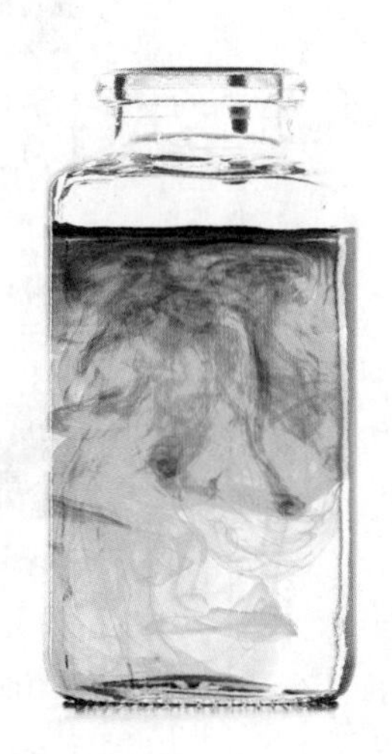

Conduction

- When a solid conductor, such as metal, is heated, the atoms increase their energy and vibrate more.
- The atoms collide with other atoms, transferring energy and causing them to vibrate more.
- The process of conduction continues until all the atoms have reached the same temperature.
- In an insulator the energy is not passed onto other atoms so the solid does not conduct the heat.
- The vibration of the particles increases as their kinetic energy increases.
- The higher the temperature, the greater the kinetic energy and so the particles vibrate faster.
- The lower the temperature, the lower the kinetic energy and the particles will vibrate more slowly.

Key Point

Kinetic energy is movement energy. The more movement, the more energy.

Temperature and Particles

- The hotter particles get, the more kinetic energy they have.
- This means particles move more and separate from each other more.
- As temperature increases, pressure will increase and the density will decrease.
- In the case of a balloon, the particles inside will increase in speed causing the **pressure** to increase, enlarging the balloon.

Movement of particles inside a warmed balloon

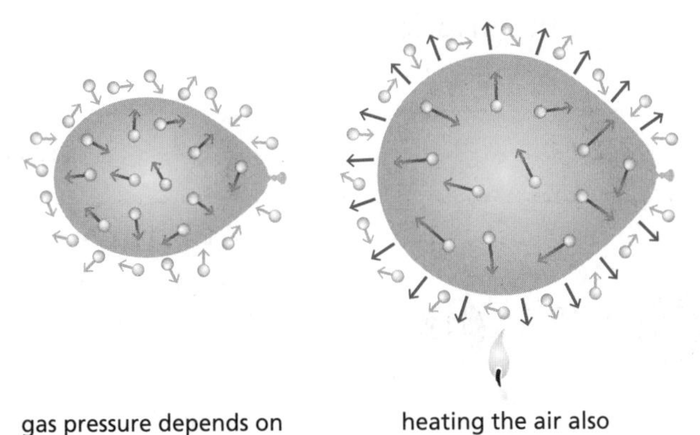

- A reaction that produces heat energy is called an exothermic reaction.
- A reaction that takes in energy from the environment is called an endothermic reaction.

Quick Test

1. What will happen to the air particles in a sealed balloon if it is heated?
2. What will happen to the air particles in a sealed balloon if it is cooled?
3. Why does ice float on water?
4. Explain how diffusion takes place.

Key Words

Brownian motion
diffusion
concentration
pressure

Explaining Chemical Changes

You must be able to:

- Apply conservation of mass to simple reactions
- Explain the combustion of fuels
- Explain the difference between a chemical and physical change
- Explain how a catalyst can make a reaction occur faster by reducing activation energy.

Chemical Reactions

- A chemical reaction involves the rearrangement of atoms from reactants to products.
- The products that are formed have the same atoms, just in different configurations.
- There is never a change in total mass in a chemical reaction.
- A **word equation** names the reactants and products formed in a reaction, for example:

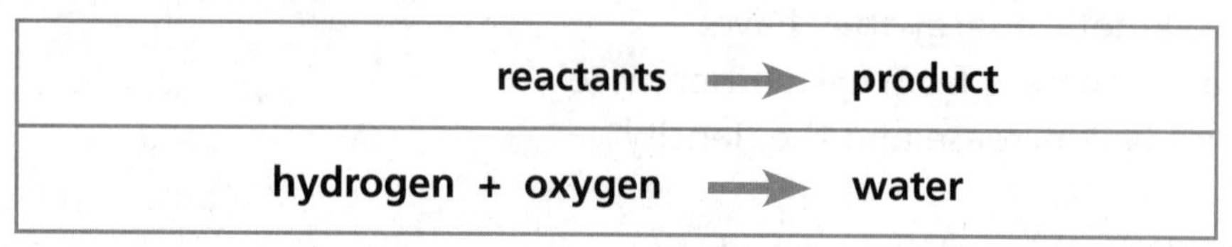

reactants	→	product
hydrogen + oxygen	→	water

- Word equations do not tell us the ratios of the molecules involved.
- **Chemical equations** show the chemical formula of the reactants and products, so that the number of atoms and ratios involved can be worked out, for example:

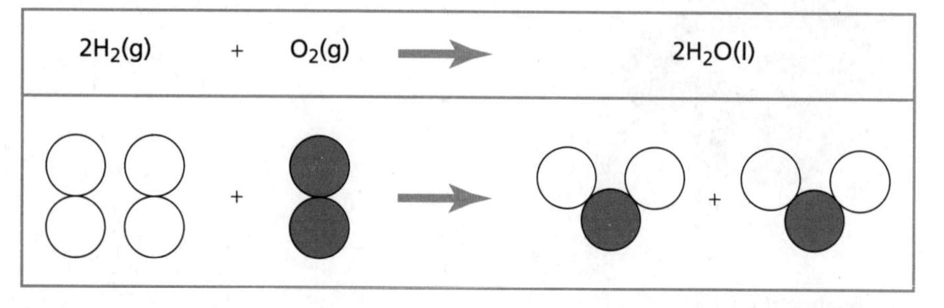

- This equation tells us that two molecules of hydrogen gas react with one molecule of oxygen gas to give two molecules of water.
- The state of each reactant is given in brackets after its chemical formula: (s) = solid, (g) = gas, (l) = liquid, (aq) = aqueous (which means it is in solution).
- In combustion, a fuel generally reacts with oxygen to produce carbon dioxide and water, and releases energy in the process. For example, burning propane gas in a camping stove:

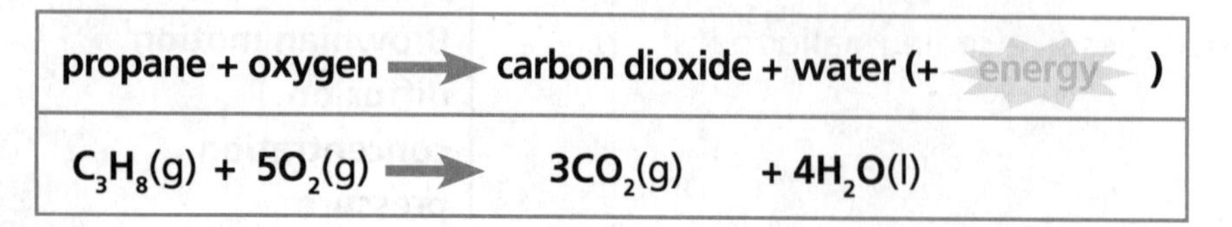

propane + oxygen	→	carbon dioxide + water (+ energy)
$C_3H_8(g) + 5O_2(g)$	→	$3CO_2(g) + 4H_2O(l)$

Key Point

The total mass in a chemical reaction doesn't change. This is conservation of mass.

Key Point

The large number in front of each molecule is a coefficient. It tells us how many molecules there are overall.

Catalysts

- A chemical reaction will only take place if a set amount of energy is provided.
- This is called the **activation energy** of the reaction.
- If the available energy is less than the activation energy, there will not be a reaction.
- A **catalyst** is a substance that reduces the activation energy, and so increases the rate of reaction.
- This means the reaction can take place with lower energy than normal.
- A catalyst is neither a reactant nor a product and is not used up in the reaction.
- The name of the catalyst is written above the arrow in the reaction to indicate that it is needed.
- Catalysts are used in chemical processes all over the world, most commonly in the exhaust systems of cars.

Catalytic convertor

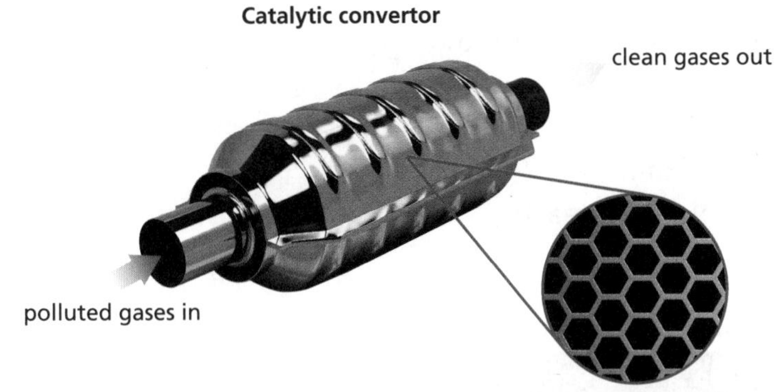

Pt catalyst
carbon monoxide + oxygen → carbon dioxide
Pt catalyst
$2CO(g) + O_2(g) \rightarrow 2CO_2(g)$

- In biology, catalysts are made of protein and are called enzymes.
- One example of an enzyme is amylase in the digestion of starch into the sugar maltose:

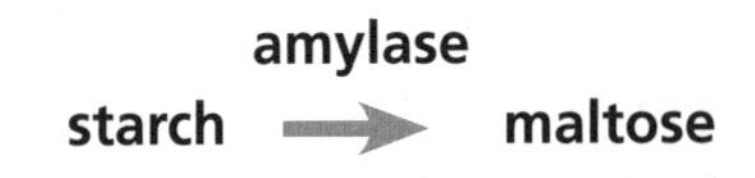

amylase
starch → maltose

Quick Test

1. What does a catalyst do?
2. Explain activation energy.
3. State the differences between word and chemical equations.
4. Write the four state symbols used in equations.

Key Point

Catalysts are never used up, so can be used again and again.

Key Words

word equation
chemical equation
activation energy
catalyst

Explaining Chemical Changes

You must be able to:

- Explain neutralisation and the use of indicators
- Use word equations to represent and/or describe the reactions of acids
- Describe and explain the uses of acids and alkalis.

Indicators

- Indicators are used to find out what type of substance a chemical is.
- Initially, scientists discovered chemicals in common plants that could change colour to indicate acid or alkali, for example, red cabbage is red in acid and blue in alkaline conditions.
- The chemical litmus also behaves as an indicator but can be incorporated into paper, so it can be transported easily.
- Universal indicator (UI) solution and paper contain a mixture of different indicators.
- These indicators change colour at a specific pH.
- pH is a measure of the strength of an acid or alkali.
- The pH scale ranges from 1 to 14 and has a colour for each pH number.

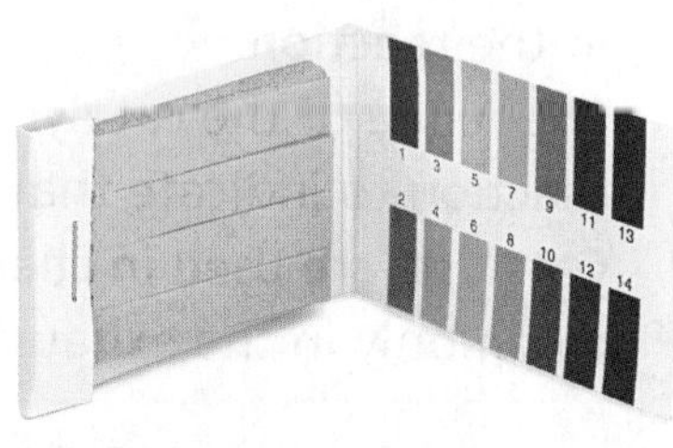

Litmus paper

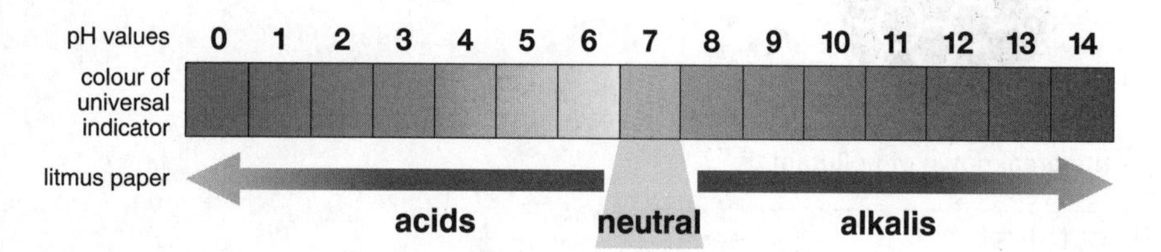

- pH probes and data loggers remove the need for indicator papers and solutions.
- They measure the pH directly and are more precise than indicator papers or solutions.

Acids and Bases

- All chemicals can be classified as being an **acid**, a **base** or are **neutral**.
- Acids are a group of chemicals that have a pH less than 7.
- An acid can chemically react with a metal to produce hydrogen:

metal + acid  **salt + hydrogen**

- The chemical opposite of an acid is a base.
- Bases are chemicals with a pH greater than 7.

Key Point

If a base dissolves, it's an **alkali**.

Name of acid	Where found	pH
Hydrochloric acid	Human stomach	1
Ethanoic acid	Vinegar	2
Citric acid	Citrus fruit	2
Sulfuric acid	Car batteries	1
Carbonic acid	Fizzy drinks	4

Citrus fruit – acid

Name of base	Where found	pH
Sodium hydroxide	Laboratories	14
Calcium carbonate	Chalk	9
Sodium bicarbonate	Bicarbonate of soda (cooking)	8
Ammonia	Hair dyes	11
Lime	Gardening products	12

Chalk – base

- When a chemical is neither an acid nor a base it is neutral.
- A neutral solution has a pH of 7.

Neutralisation

- When an acid and a base are mixed together, they react.
- If an acid is reacted with a base there will come a point where a salt and water are made and no more acid or base exists.
- At this point the mixture will be neutral and have a pH of 7.
- The whole process is called **neutralisation**.

acid + metal oxide	→	**salt + water**
acid + metal hydroxide	→	**salt + water**
acid + metal carbonates	→	**salt + water + carbon dioxide**

Key Point

In all neutralisation reactions, water is made.

Quick Test

1. What can you deduce if a chemical is pH 5?
2. Explain neutralisation.
3. Write the word equation for the reaction of acid and metal.
4. Why is a pH probe and a data logger better to measure pH than UI paper?

Key Words

acid
base
neutral
alkali
neutralisation

Mixing, Dissolving and Separating

1. Anna and Kala are carrying out chromatography of ink, as they believe a cheque has been forged. They are going to test to see whether the ink on the cheque is the same as the ink of the suspect's pen. The results are shown in the chromatogram below.

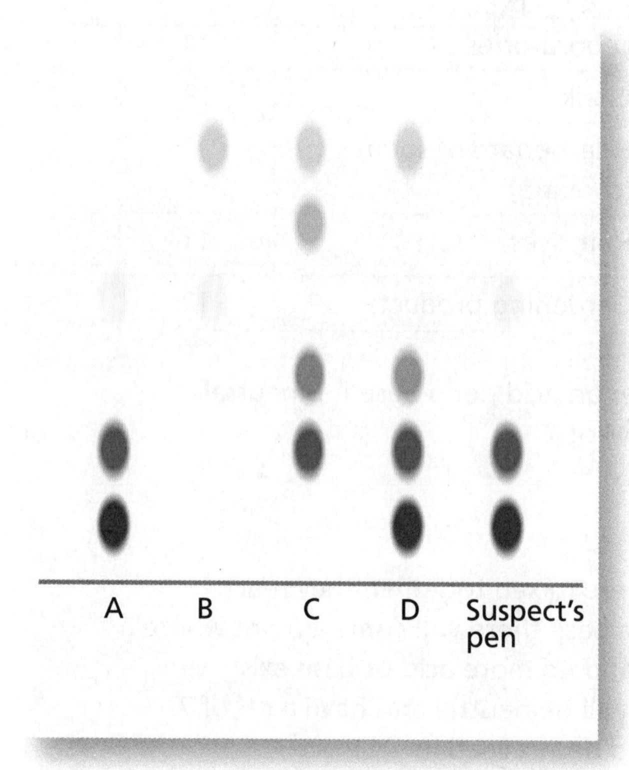

a) How many different pigments are in sample A and in sample B? [2]

b) Which of the samples matches the ink of the suspect's pen? [1]

c) Explain why was the starting line on the chromatogram was drawn using a pencil. [2]

2. Karim is investigating how chalk (calcium carbonate) reacts when heated. He heats the chalk for a minute at a time and then measures its mass.

He notices that the mass decreases.

a) What is the name given to the chemical reaction Karim is observing? [1]

b) Complete the word equation for the reaction given below:

calcium carbonate → .... + ... [2]

3. Write the chemical equation for the reaction of carbon with oxygen. [3]

Elements, Compounds and Reactions

1 The diagram shows a model of a chemical reaction.

a) What feature of the diagram indicates that a chemical reaction has taken place? [2]

b) Substance X is hydrogen. Suggest what substances Y and Z could be. [2]

c) Suggest how the diagram indicates that mass has been conserved in the reaction. [1]

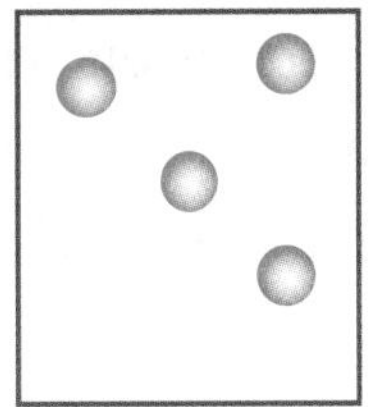
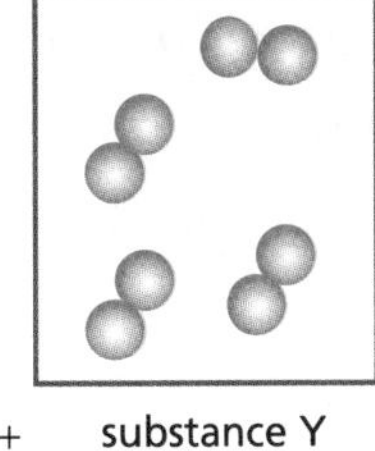
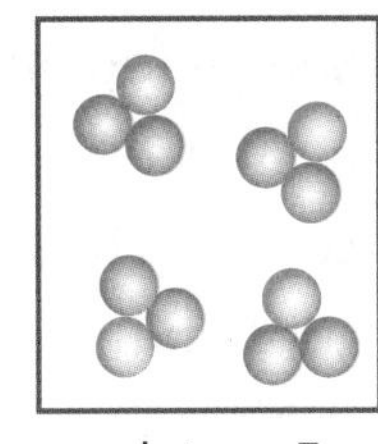

substance X + substance Y → substance Z

2 Niamh is carrying out some experiments to observe what happens when different metals are added to different acids.

Her table has some gaps. What are the reactants and products given by letters A–E?

Metal	Name of acid	Salt formed	Gas formed
Zinc	**A**	Zinc sulfate	Hydrogen
Magnesium	Nitric acid	**B**	Hydrogen
Potassium	**C**	Potassium chloride	**D**
E	Nitric acid	Lithium nitrate	Hydrogen

[4]

3 For each change given below, decide whether it is a physical change (P) or a chemical change (C).

a) melting chocolate

b) a burning firework

c) the smell of perfume diffusing across a room

d) jelly setting in a dish

e) setting off a CO_2 fire extinguisher [1]

4 Which of the following chemical equations is correct?

a) $Mg(s) + O_2(g) \rightarrow 2MgO(s)$

b) $2Mg(s) + O(g) \rightarrow MgO(s)$

c) $2Mg(s) + O_2(g) \rightarrow 2MgO(s)$

d) $Mg(s) + O(g) \rightarrow MgO(s)$ [1]

Explaining Physical Changes

1. Draw the particles in a solid, liquid and a gas. [3]

2. Sally is looking at a blue coloured gel air freshener, which sublimes. She cuts a piece of the air freshener and puts it into a small beaker. The beaker is placed into a larger beaker which contains hot water. On the top of the beaker containing air freshener, she places a beaker containing ice.

   a) What would Sally see after a few minutes on the underside of the beaker containing ice? [1]

   b) Suggest why this happens. [1]

3. What is meant by the term Brownian motion? [2]

4. An iceberg floats on water. Explain what properties of water enable ice to float. [3]

5. Claude is holding a CO_2 fire extinguisher. When he sets the extinguisher off, a valve opens releasing CO_2 gas.

   When a handkerchief is placed over the end of the extinguisher, solid CO_2 starts to build up.

   Which of the following best explains why this happens? Tick the correct box.

   a) The CO_2 particles break up into C and O_2 ☐

   b) The CO_2 gas particles are slowed down rapidly causing a solid to form ☐

   c) The CO_2 gas particles slow down to form a liquid and then a solid ☐

   d) The handkerchief cools the CO_2 down into a solid ☐ [1]

Explaining Chemical Changes

1. Mark is reacting copper metal with oxygen gas. He draws the atoms involved in the reaction.

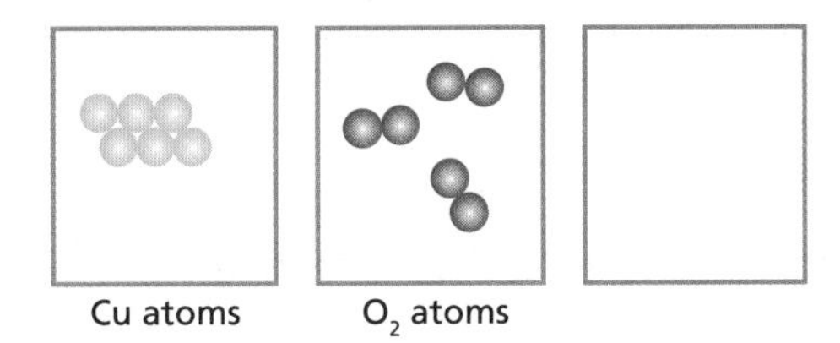

   a) Draw what should be in the last box. [2]

   b) What is the name of the product in this reaction? [1]

2. What does a catalyst do in a chemical reaction? Choose the best answer.

   a) It slows a chemical reaction down

   b) It makes chemical reactions happen

   c) It lowers the activation energy

   d) It lowers the kinetic energy [1]

3. Complete the following:

   metal + acid → .... + ... [1]

4. Peter is using universal indicator to identify acids and bases.

   Suggest what colour UI would change to for each of the examples below:

   a) car battery acid

   b) juice of a lemon

   c) tap water

   d) toothpaste [4]

5. What are the products of the following reactions?

   a) magnesium + hydrochloric acid → [1]

   b) copper oxide + nitric acid → [1]

   c) vanadium carbonate + sulfuric acid → [1]

Obtaining Useful Materials

You must be able to:

- Use the reactivity series to determine whether reactions are possible
- Describe and explain how carbon is used to extract metals.

The Reactivity Series

- The metals in the periodic table all exhibit different levels of **reactivity**.
- By comparing their reactions it is possible to sort them into a reactivity series.
- The reactivity series shows the order of reactivity, from most reactive to least reactive.

Most reactive

Potassium
Sodium
Calcium
Magnesium
Aluminium
Carbon
Zinc
Iron
Tin
Lead
Copper
Silver
Gold
Platinum

Least reactive

- By comparing metals on the reactivity series, chemists can predict whether a chemical reaction may happen and, if it does, how vigorous the reaction may be.
- The reactivity series also includes the element carbon.
- Carbon can have different forms and properties, such as diamond and graphite.
- Carbon is not a metal, but when it is in the form of graphite it can conduct electricity, a property shared with the metals.

Key Point

All metals could be included in a reactivity series. The diagram here shows only a few major metals, as well as carbon.

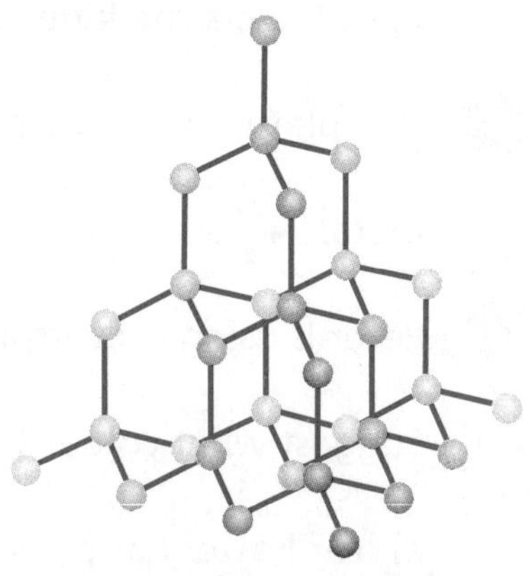

Structure of diamond
Diamond is very strong

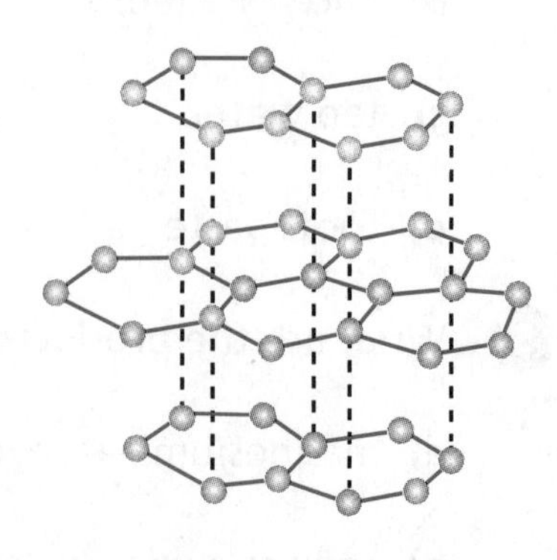

Structure of graphite
Graphite conducts electricity

Displacement Reactions

- When a metal, or carbon, comes into contact with a metal that is in a compound there may be a reaction:
  - If the metal in the compound is higher in the reactivity series than the introduced metal, or carbon, no reaction will take place.
  - If the metal in the compound is lower in the reactivity series than the introduced metal or carbon then the introduced metal will **displace** the metal in the compound.
  - The greater the difference in positions in the reactivity series, the faster and more vigorous the reaction.
- For example, if iron metal was added to copper sulfate solution:

Key Point

Remember that, as well as new products being formed, there will often be visible changes as a result of the reactants disappearing.

Iron nail in copper sulfate solution

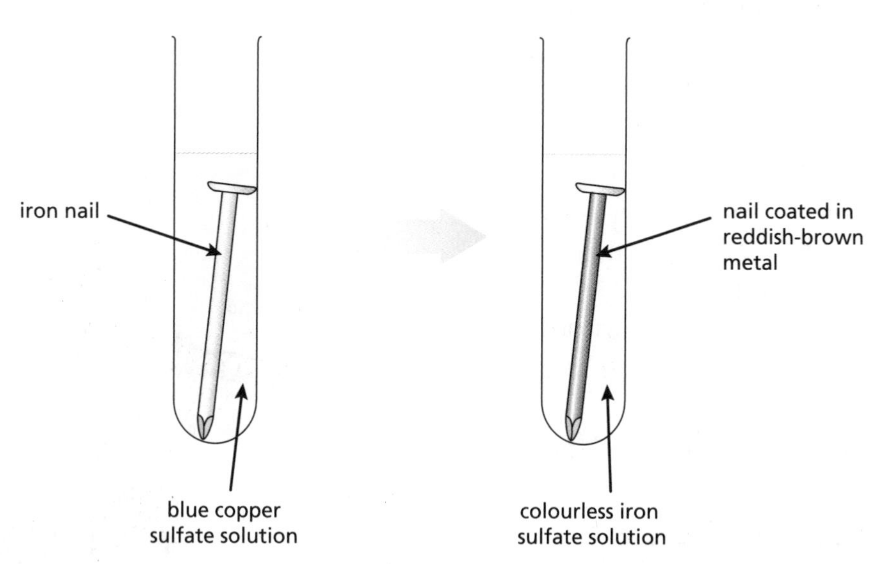

Iron is higher than copper on the reactivity series. It will therefore displace the copper and form iron sulfate. Iron is a silver coloured metal. It will slowly dissolve.
The blue copper sulfate solution will lose its colour as the copper is displaced and colourless iron sulfate is formed. The copper, which was previously in solution, will form a reddish-brown solid.

- The reaction can be written as:

iron + copper sulfate	→	iron sulfate + copper
$Fe(s) + CuSO_4(aq)$	→	$FeSO_4(aq) + Cu(s)$

- If copper were added to iron sulfate solution, there would be no reaction as copper is lower in the reactivity series than iron and will not be able to displace it.

Quick Test

1. Could copper displace magnesium from magnesium nitrate?
2. Which would be faster at displacing zinc – iron or aluminium?
3. Suggest a metal to displace sodium from sodium chloride.
4. Explain what is meant by the 'reactivity series'.

Key Words

reactivity
displace

Chemistry

Obtaining Useful Materials

You must be able to:
- Describe and explain the use of carbon in the extraction of metals
- Describe the characteristics and uses of ceramics, composites and polymers.

Extracting Metals

- Many of the first metals discovered by humans were low on the reactivity series.
- As they were unreactive it meant that the metals could be found in their pure form, e.g. gold and silver, and were not in a compound.
- Other, more reactive, metals tended to have already reacted with other elements, such as oxygen, to form compounds.
- One of the biggest milestones in human history was stumbling upon the displacement reaction using carbon that can be used to purify iron ore:
  - Iron is heated in a furnace with carbon and limestone
  - Carbon is higher in the reactivity series than iron so displaces the iron
  - Molten iron is formed and the carbon joins on to the oxygen forming carbon dioxide.

carbon + iron oxide	→	carbon dioxide + iron
C(s) + 2FeO(s)	→	CO_2(g) + 2Fe(l)

Key Point

All metals below carbon in the reactivity series can be extracted using this method.

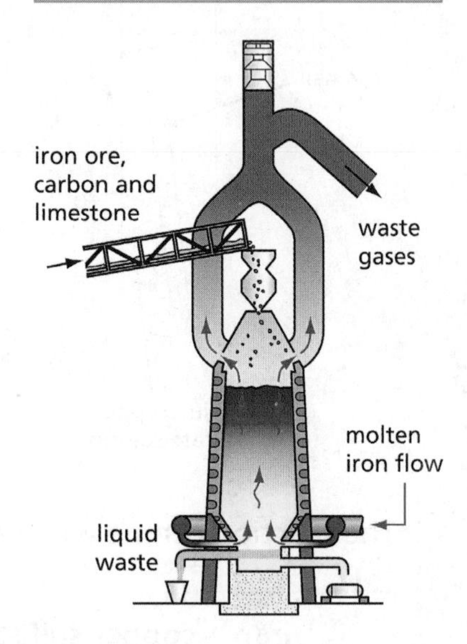

Ceramics

- Ceramics are made from heating non-metallic materials at high temperatures.
- The properties of the ceramic material differ from the initial material, e.g. ceramic pots are very different to the clay used to make them.
- This is because the high temperatures cause crystals to form on cooling.
- By controlling the speed of cooling, different sized crystals can be made.
- Rapid cooling causes small crystals and slow cooling causes large crystals.

- If other minerals are added to glass when it is formed, glass ceramics can be made.
- Although they are often brittle, these can tolerate very high temperatures, so are often used as cookware (e.g. Pyrex) or in laboratories.

Key Point

Materials such as clay need to be heated in an oven to become a ceramic. This is an irreversible process.

Polymers

- It is possible to join small molecules (**monomers**) together in long chains.
- The chained molecule consisting of repeating monomer units is called a **polymer**.
- Polymers are extremely useful because they can be used to create different materials, e.g. the monomer ethene can be made into poly(ethene).
- Poly(ethene) is used to make plastic bags and bottles.

$$n\,\mathrm{CH_2{=}CH_2} \rightarrow \left(\!-\mathrm{CH_2{-}CH_2}-\!\right)_n$$

ethene → poly(ethene)

n means a large number of, or many

Composites

- A **composite** is a material that is made from two or more different materials bonded (joined) together.
- The new composite material has different characteristics to those of the starting materials.
- Concrete, first made by the Romans, is made from mixing cement with different stones. The resulting mixture is far stronger than the cement or the stones alone.
- Carbon fibre is a very light and exceptionally strong composite formed from sheets of carbon fibre bonded together with a resin.
- Carbon fibre materials are used when strength is needed with low weight; for example, helicopter rotor blades, airplanes and kayaks.

Quick Test

1. Why were gold and silver amongst the first elements discovered?
2. How would you create small crystals in clay?
3. What is a composite?
4. Explain how to make a polymer.

Key Words

ceramic
monomer
polymer
composite

Chemistry

Using our Earth Sustainably

You must be able to:
- Describe the structure of the Earth
- Explain rock formation and the rock cycle
- Describe the composition of the Earth and its atmosphere.

The Structure of the Earth

- The Earth is a planet in orbit around the Sun.
- All of the chemicals used in industry come from the Earth.
- Rocks make up the solid crust of the Earth.
- The rocks in the crust contain chemical compounds and elements that can be extracted and used.

Structure of the Earth

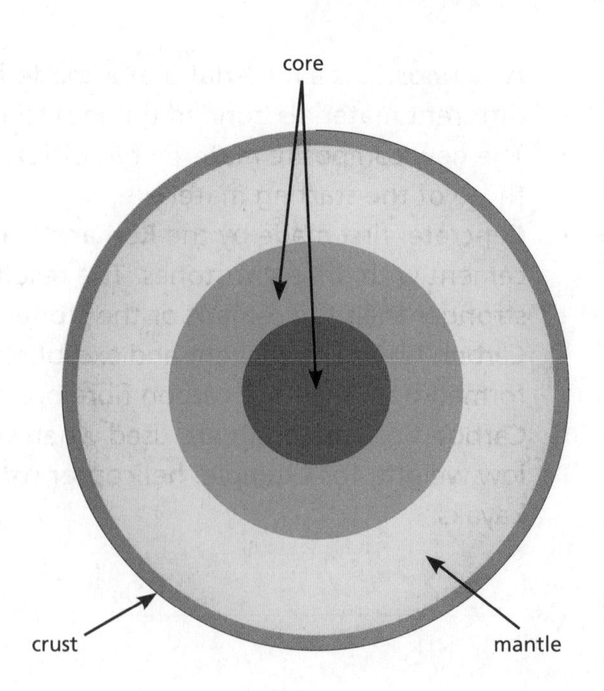

The Rock Cycle

- There are three types of rocks, classified according to how they formed.
- **Igneous** rock is formed from cooled magma.
  - If the magma cools rapidly, the igneous rock formed is dense and has small crystals, e.g. granite
  - If the magma cools slowly then the igneous rock will be less dense and have large crystals, e.g. basalt.
- When rocks have been subjected to weathering and erosion, pieces of rock break off.
- The sediments formed eventually settle and are subjected to large pressures from the rock above.
- When material deposited in this way forms rock it is called **sedimentary** rock.
- Any material that can be deposited can lead to sedimentary rock, e.g. eroded rock material, or the calcite skeletons of microorganisms that lived in the sea.
- Due to the movement of the tectonic plates, rock that was on or near the surface can be moved closer to the Earth's core.
- Subjected to intense heat and pressure, the structure of the rock alters.
- The altered rock becomes **metamorphic** rock.

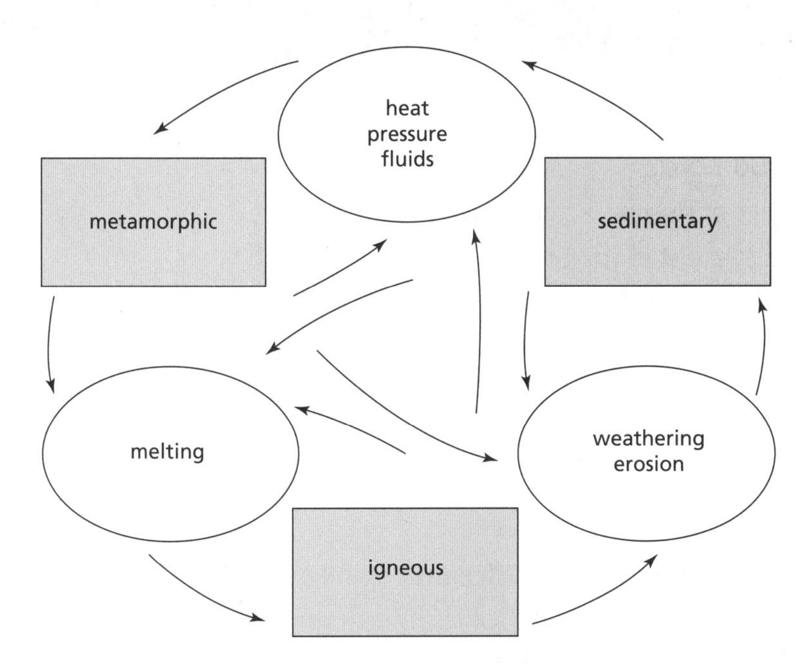

Sedimentary rock

Igneous rock

Metamorphic rock

Composition of the Earth

- The chemicals in the Earth's crust tend to be insoluble compounds. If they were soluble, the ground would dissolve whenever it rained!

Element	Abundance (%)
Oxygen	46
Silicon	28
Aluminium	0.8
Iron	0.5
Other elements	24.7

- The atmosphere of the Earth contains very light compounds that exist as gases, as well as very small solid particles that are light enough to float in the air.

Element	Abundance (%)
Nitrogen	78
Oxygen	21
Argon	0.9
Carbon dioxide	0.04
Other trace gases	0.06

Key Point

The composition of elements is very different in different parts of the Earth. It is important to specify which part you are describing.

Quick Test

1. What are the three types of rock?
2. Why do chemicals in the lithosphere tend to be insoluble?
3. Draw the rock cycle.

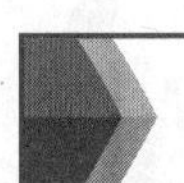

Key Words

igneous
sedimentary
metamorphic

Chemistry

Using our Earth Sustainably

You must be able to:

- Describe the carbon cycle
- Explain the impact of human activity on the atmosphere
- Describe what is present in the atmosphere and how it has changed over time
- Suggest why the Earth is a source of limited resources.

The Carbon Cycle

- The element carbon (C) is common on Earth.
- Carbon is reactive and can form up to four bonds with different elements, often forming chains.
- The different reactions that carbon takes part in mean that carbon atoms move through the **carbon cycle**.
- Changes to parts of the cycle will have an impact on other parts of the cycle.

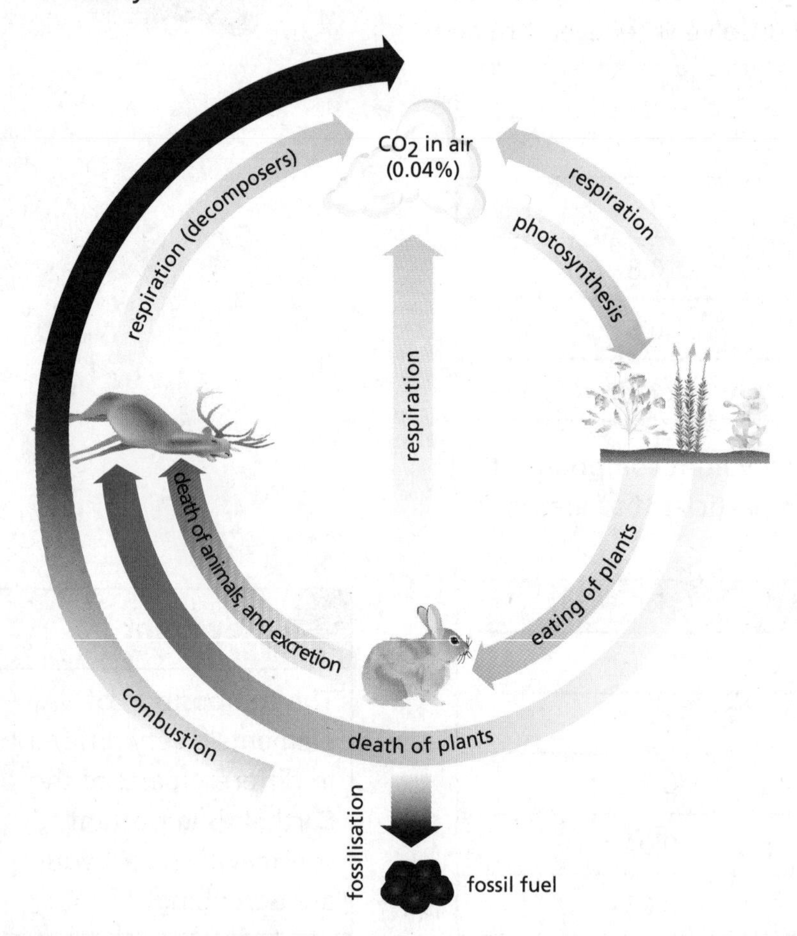

Human Activity and Climate Change

- Human activity can alter the balance in the carbon cycle.
- By removing and burning **fossil fuels** (coal, oil and natural gas), carbon that was trapped for millions of years is now released into the atmosphere as carbon dioxide.

Key Point

Levels of CO_2 are globally the highest they have been for the past 3 million years.

- This carbon dioxide acts as a greenhouse gas, trapping heat from the Sun.
- The more carbon dioxide released, the hotter the planet becomes – the consequence of this is climate change.
- In different parts of the world, the oceans get warmer.
- This causes changes to currents and wind patterns, and unpredictable and extreme weather patterns can occur.
- Humans produce carbon dioxide through a number of activities:
  - factories or transport (e.g. planes, cars) using fossil fuels
  - cutting down and burning forests.
- Scientific **consensus** indicates that the amount of carbon dioxide in the atmosphere needs to be reduced.
- Before industrialisation in the 1800s the amount of carbon dioxide in the atmosphere was 0.028%.
- In 2013 the level reached 0.04% for the first time in the past 3 million years.

Key Point

Although the media often show both sides of an argument, it is important to recognise when one side has far more scientific evidence than another.

Limited Resources

- Although the planet looks enormous to us and seems to have an endless supply of resources, this is not true:
  - Many elements that we find most useful are also rare
  - Many of the components used in mobile phones and tablets are made using rare earth metals
  - The majority of the energy we use comes from non-renewable sources, such as coal, oil and natural gas
  - Most items in households of the Western world are made from the products of crude oil.
- The finite reserves of oil are being used up, and since oil takes millions of years to form, we cannot make more. This means that we need to recycle items.
- **Recycling** involves extracting parts of a used product and making them available for other processes or products, e.g. recycling paper involves collecting the used paper, sorting it and then treating it to make recycled paper.
- Of course, recycling uses energy to extract the materials.
- Manufacturers are increasingly being required by law to make the extraction and recycling of materials easy.

Quick Test

1. What do scientists believe is causing CO_2 levels to increase?
2. What is recycling?
3. Draw the carbon cycle.
4. How are humans affecting the Earth's atmosphere?

Key Words

carbon cycle
fossil fuels
consensus
recycling

Explaining Physical Changes

1 The diagram below shows the arrangement of atoms in four different substances.

A

B

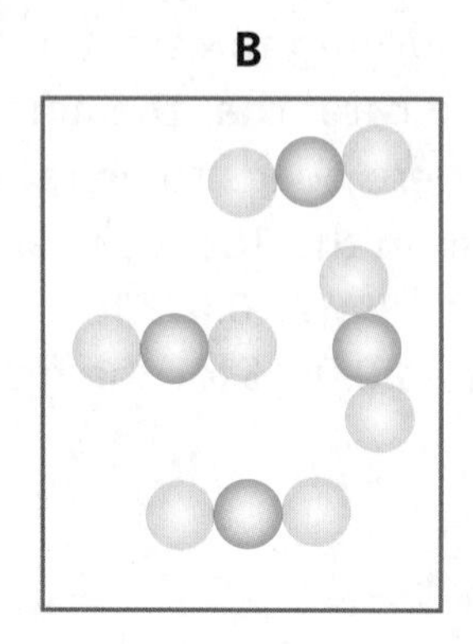

C

D

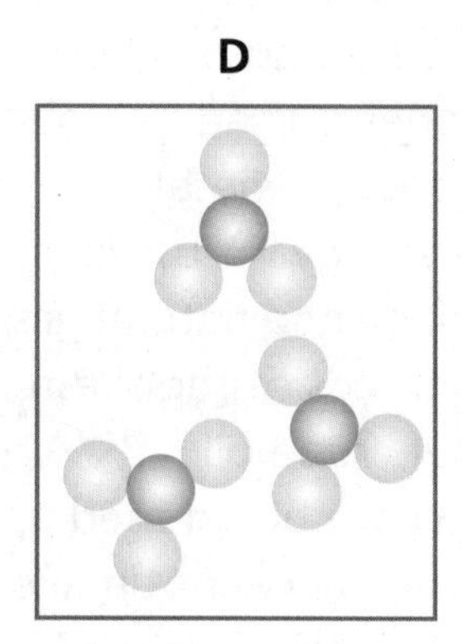

a) Which box has a substance that is a mixture of compounds? [1]

b) Which box has a substance that could be carbon dioxide? [1]

c) Which box has a substance that is a pure element? [1]

2 It is winter and Sylvia has gone to look at her fishpond. The water pump is switched off. There is a layer of ice over the top of the pond. Despite the ice, the fish in the pond are alive and are able to swim.

a) Explain what property, unique to water, has enabled the fish in the pond to survive. [3]

b) Sylvia's sister Hilary suggests breaking the ice and switching the water pump back on.

Explain why this could lead to the death of the fish in the pond. [3]

Explaining Chemical Changes

1. a) Write the word equation for the reaction of sodium and chlorine. [2]

   b) Write the chemical equation for the reaction of sodium, Na, and chlorine, Cl. [2]

2. Hydrogen peroxide is a chemical that acts as a bleach.

   To decompose hydrogen peroxide into water and oxygen the chemical manganese dioxide needs to be present.

   a) Write the word equation for the decomposition of hydrogen peroxide. [2]

   b) Manganese dioxide is neither a reactant nor a product. What is the chemical name for this type of chemical? [1]

3. In 2012 Feliz Baumgartner floated in a helium balloon until he reached a height of 39 km before jumping out.

   a) The diagram below shows a balloon that is floating.

   Copy the diagram and draw arrows inside the balloon to represent the pressure of the gas inside and arrows representing the air pressure outside the balloon. [2]

   b) Draw the arrangement of particles before and after the air inside the balloon was heated. [1]

Obtaining Useful Materials

1. The list below is a shortened reactivity series. Use it to answer the following questions.

potassium
sodium
calcium
aluminium
carbon
iron
tin

a) Which of the metals can be extracted from their ores by reacting with carbon? [2]

b) What would be the products of the displacement reaction between carbon and tin carbonate?

Complete the word equation.

carbon + tin carbonate → .... + ...

[2]

2. What are the small units called that make up a polymer? [1]

3. The volcanic rock basalt is one of the most common rocks on the Earth. It has small crystals that can only be seen clearly using a microscope.

Suggest how quickly basalt cooled and why it would lead to crystals of a small size. [2]

4. Colin is studying the architecture of the Roman Civilisation. He reads that the Romans were the first to make concrete. Concrete is made from stones and cement.

Explain why concrete is a composite material and suggest why the Romans chose to use concrete in their buildings, rather than cement or stones on their own. [2]

Using our Earth Sustainably

1. The Earth consists of layers. Copy the picture below and add labels to identify each layer. [3]

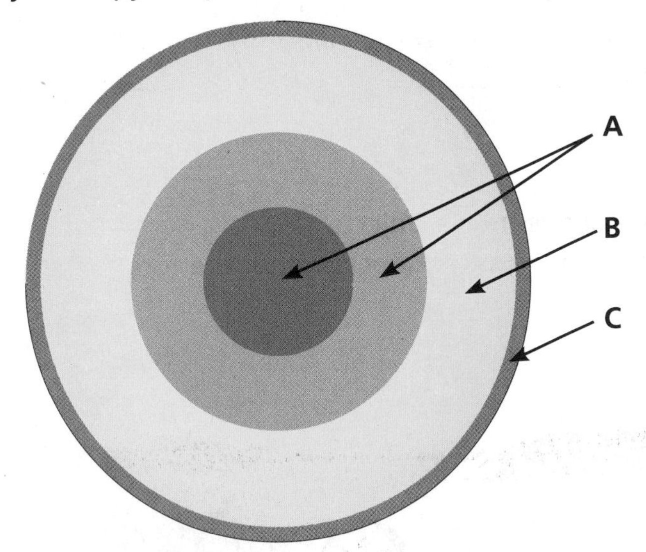

2. The rock cycle below is missing some labels. Copy the diagram and fill in the gaps. [5]

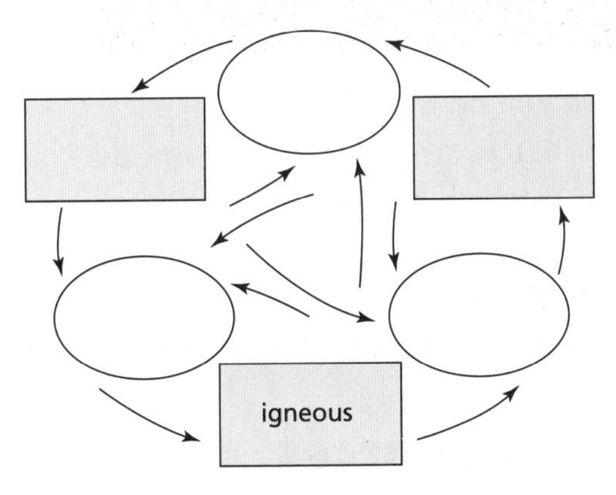

3. The IUPCC has provided evidence that man's activities are a direct cause of climate change.

   a) What is climate change? [1]

   b) Which of the following are direct causes of climate change?

   i) using nuclear power stations

   ii) manufacturing plastics from crude oil

   iii) geothermal energy

   iv) petrol powered cars [2]

4. Explain why it is important to recycle resources. [2]

Forces and their Effects

You must be able to:

- Describe some causes and effects of forces
- Explain how objects can be affected by forces
- Use force arrows to show forces acting on objects
- Explain the concept of moments
- Describe and explain Hooke's Law.

What are Forces?

- **Forces** are pushes or pulls between two objects.
- Force arrows can be drawn to show the direction that the force is acting in:

- When forces act on an object there will be a consequence:
  - The object can become deformed, e.g. stretched or squashed
  - The object can warm up due to rubbing and the friction between the surfaces
  - The object can be pushed out of the way
  - The object can provide resistance to the motion of water or air.

Key Point

You should clearly identify the direction forces act in using arrows.

Balanced and Unbalanced Forces

- Forces act in opposite pairs and force arrows are drawn to show this.
- The forces acting on an object can either be:

Balanced	Unbalanced
This means that there will be no change in the object.	This means that there will be a change in the object.
The upward push of the shelf is balanced by the downward pull of the earth.	The driving force of the cycle is greater than the air resistance - the force is unbalanced. The cyclist is accelerating.

Measuring Forces

- Forces are measured in **Newtons** (N).
- The stronger the force, the greater the value in Newtons.
- Forces can be measured using a Newton meter.
- The size of a force can also be measured by the size of the force's effect.
- When an object is stretched or squashed the length can be measured.
- The greater the force applied, the greater the stretch or squashing force.
- There is a linear relationship between the size of the force and the resulting stretch (i.e. they change in proportion to each other).
- This means that for every unit of force, there will be the same effect on the stretch (e.g. 1 N causes 3 cm stretch, 2 N causes 6 cm stretch, 10 N causes 30 cm, etc.)

Key Point

The unit for force, the Newton, must ALWAYS be written with a capital N.

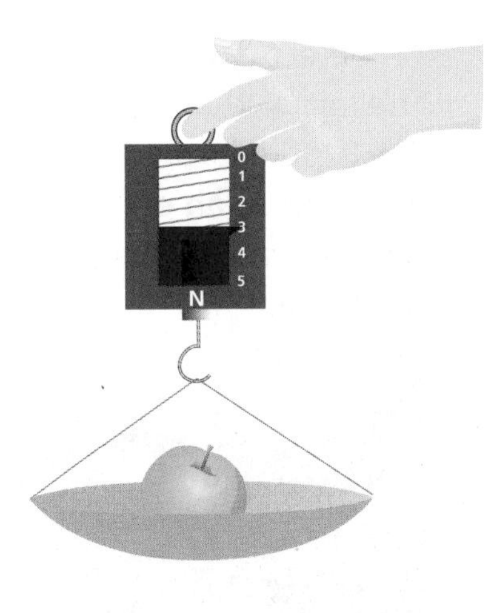

Pivots and Moments

- A **pivot** is used to turn things.
- A **moment** is the turning effect of a force, and has the unit Nm.
- moment = force x distance from pivot.

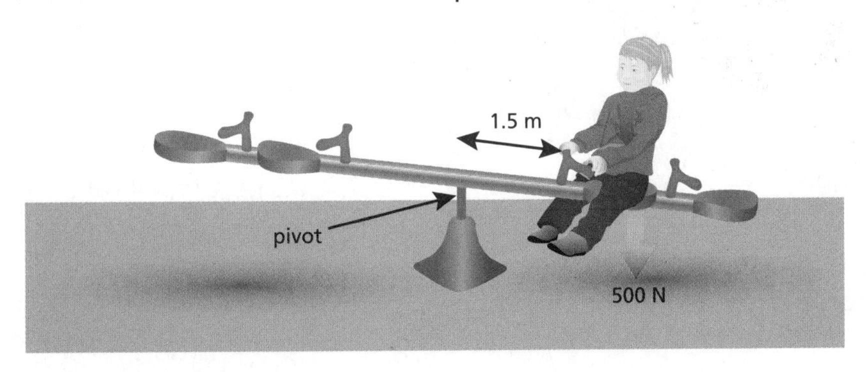

- The further away from the pivot, the greater the moment.
- This is why a long-handled screwdriver can be used to open a tin of paint.

Hooke's Law

- Hooke's Law states that the stretch of a spring will be directly proportional to the force applied (i.e. there is a linear relationship between them).

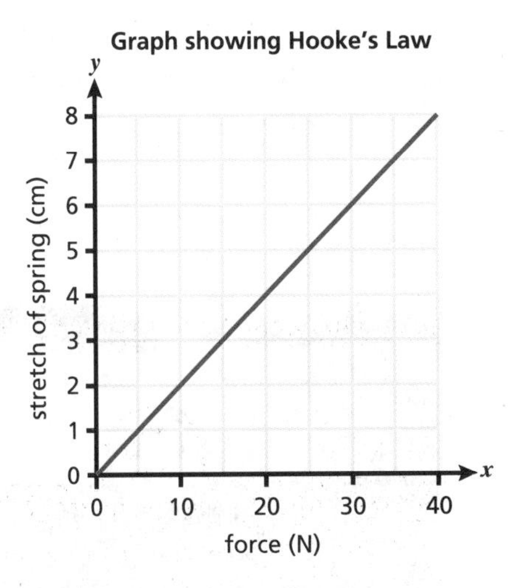

Quick Test

1. How do we describe a pair of forces if one force is larger than the other?
2. What is the unit for force?
3. What do force arrows tell us?
4. What is meant by the term moment?

Key Words

force
Newton
pivot
moment

Forces and their Effects

You must be able to:

- Explain simple effects of forces on speed and direction
- Calculate speed from distance and time
- Explain how work is done when a force acts on an object and changes it
- Explain how deformation involves work.

Forces and Motion

- An object will move in the direction a force is applied.
- The larger the force, the faster the object will start to move.
- Applying a large enough counter force in the opposite direction to an object's movement will cause the object to slow down and stop, for example, a football net catching a football.

Key Point

When forces are balanced an object will continue at the same speed. An object that isn't moving has a speed of 0 m/s. If it isn't already moving, it will continue to not move.

- To change an object's **speed**, a force has to be applied.
- For an object to speed up, more force is needed in the direction of movement.

- For an object to slow down, the force has to be in the opposite direction to the direction of movement.

Car travelling on tarmac at constant speed 70 mph.

There is greater friction between thegravel and the car tyres.
To travel at 70 mph, more force has to be producedby the engine.

Speed, Distance and Time

- Speed is a measure of the distance that an object travels in a given time.
- speed = distance / time
- The faster the speed, the further the object travels in a set period of time.
- If an athlete ran 100 m in 10 s then:

  speed = 100 m ÷ 10 s

  = 10 m/s
- The magnitude of the change in speed of an object depends on the size and direction of the force that is applied.

Work and Energy Changes

- In science, **work** is done if a force has acted on an object and there has been a change in that object.
- So, if an object did not move at all, no work will have been done.
- But, if the object moved faster or slowed down, then work will have been done.
- If the object becomes deformed (squashed or stretched) then work will also have been done.
- Work is measured in Newton metres (Nm), or **Joules** (J).

Key Point

Work is only done if the object changes speed or shape.

Quick Test

1. Give the unit for speed.
2. What is meant by the term work?
3. A car drives 48 km in 40 min. What is its speed?

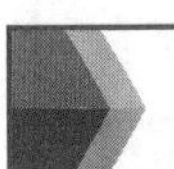

Key Words

speed
work
Joule

Exploring Contact and Non-Contact Forces

You must be able to:

- Explain non-contact forces between magnets and static electricity
- Explain electrostatic attraction and repulsion
- Describe the effects of gravity across space.

Non-Contact Forces

- A contact force is applied by one object touching another.
- Forces can also act over a distance. These are called non contact forces, e.g. magnetism, static electricity and gravity.

Magnetism

- Magnets have two poles, North and South.
- The opposite poles of a magnet **attract**.
- The same poles of a magnet **repel**.

Key Point

Always remember to write both poles on diagrams of magnets.

Magnets attracting and repelling

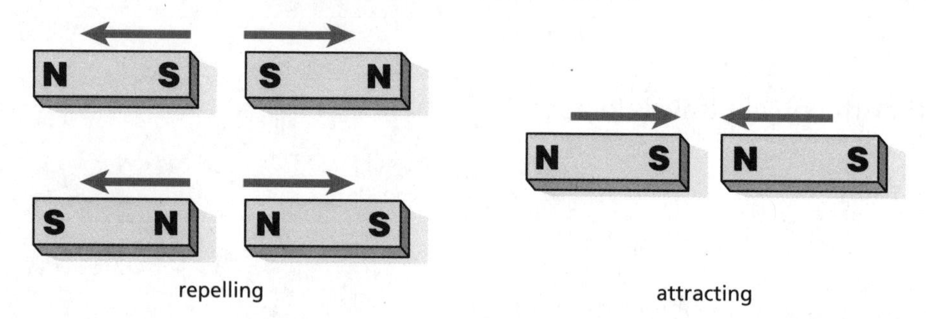

Static Electricity

- Static electricity occurs when an object gains or loses electrons (negative charge).
- If the object gains electrons it becomes negative; if it loses electrons it becomes positive.
- If two objects with the same charge are brought together, they will repel each other.
- If an object with a negative charge is brought near an object with positive charge, they will be attracted.

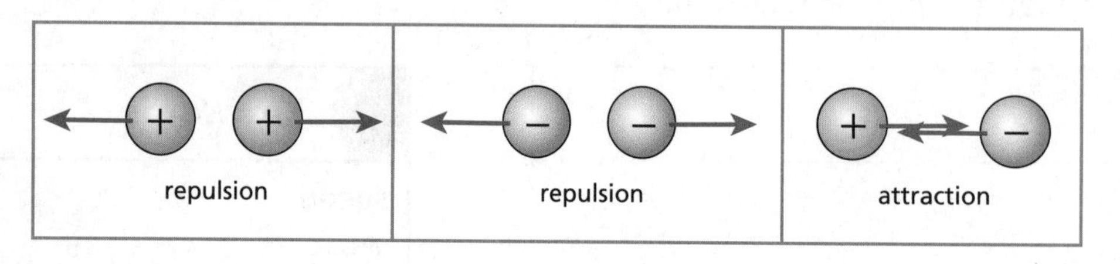

Electric Fields

- When an electric current passes through a wire it produces an **electric field.**
- Electric fields act across the spaces between objects that are not in contact with one another.

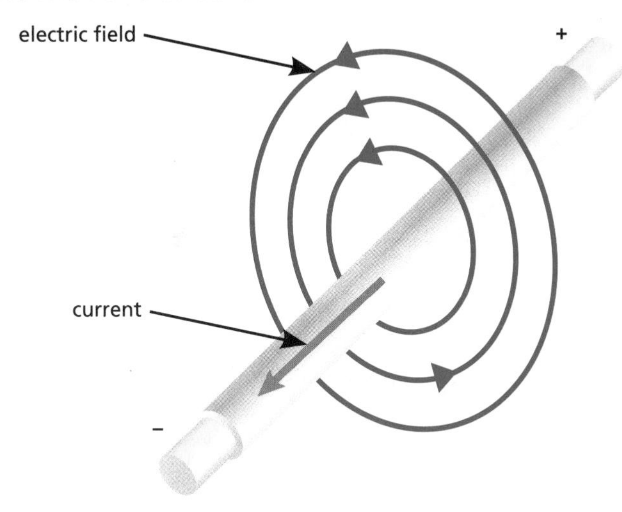

Gravity

- Gravity is a force exerted by one object on another when they are near each other.
- On the Earth everything is pulled to the Earth's centre.
- The Earth has a gravitational field strength of 10 N per kg. This means every kilogram on Earth has a force of 10 N acting on it.
- Every object with mass has a **weight**, measured in N.
- weight = mass x gravitational field strength
- Mass is the measure of all the matter in an object, and has the units kg.
- On other planets and moons the gravitational field strength will be different. The weight of an object will therefore be different on each planet, but the mass will stay the same.
- All objects with mass have a gravitational pull, even people. As the earth is so massive, we do not notice these gravitational pulls.
- The further away you are from the centre of mass causing the gravitational field, the weaker the gravitational force. For example, a person standing on the top of Mount Everest will experience slightly less gravitational pull than a person standing in Trafalgar Square, London.

Key Point

Even in space there is still gravity. When someone is 'weightless' it is because everything is falling at the same time.

Key Point

Gravitational field strength is measured in N/kg.

Quick Test

1. What is the difference between weight and mass?
2. Why does weight decrease with distance from the Earth?
3. What forms around a wire when electric current flows through it?
4. Draw two magnets repelling and attracting.

Key Words

attract
repel
electric field
weight

Physics

Exploring Contact and Non-Contact Forces

You must be able to:

- Explain how pressure acts in the atmosphere, in liquids and in solids
- Explain pressure as the effect of force over area
- Explain why objects float and sink.

Pressure

- Forces can act over an area, in all directions.
- **pressure** = force / area
- Pressure is measured in Newtons per metre squared (N/m^2).
- A person wearing snowshoes exerts less pressure on the ground than a person of the same mass wearing ice skates.

Key Point

Make sure that the unit is written N/m^2 and not Nm, which is the unit for the moment of a force.

Atmospheric Pressure

- The atmosphere on the Earth is exerting a pressure on all objects on the surface.
- As an object gets higher, atmospheric pressure reduces, because there is less atmosphere pushing downwards.

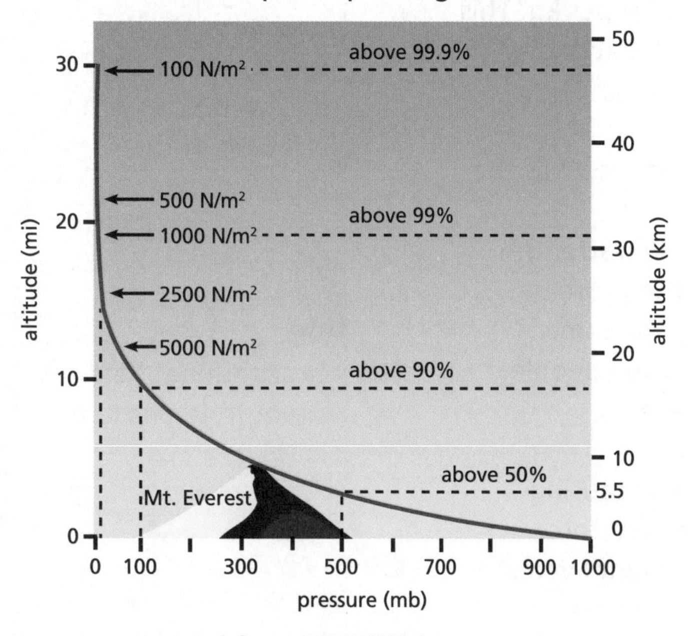

pressure at surface = 101,000 N/m^2

Pressure in Liquids

- With liquids, pressure increases with depth.
- The deeper an object gets, the greater the force acting on it due to the weight of the liquid above.
- The pressure at the top of the bottle is less than the pressure towards the bottom. Water pours out of the bottom hole much faster than the hole at the top.

Water coming out of 3 holes in a bottle

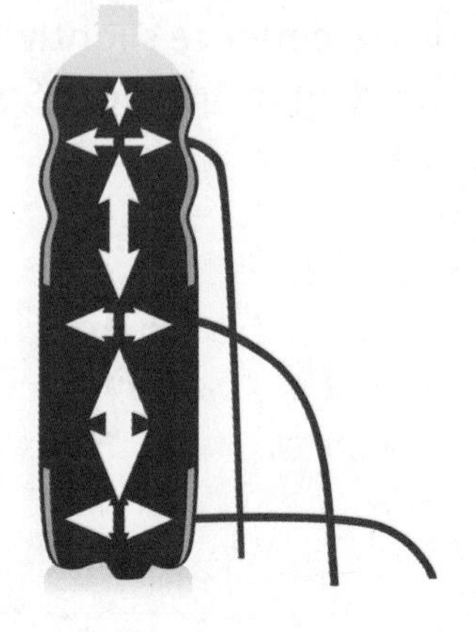

Upthrust

- When an object is placed into water, the water exerts a force in the opposite direction to the weight of the object.
- The term for this is **upthrust**.
- The more dissolved salts in the water, the greater the upthrust.
- Many ships have a series of lines, called **plimsoll lines**, painted on the hull to show how deep the ship will sink in different waters. In fresh water a ship will sink lower whilst in very salty water, such as in the Dead Sea, the ship will sink less due to the increased upthrust.

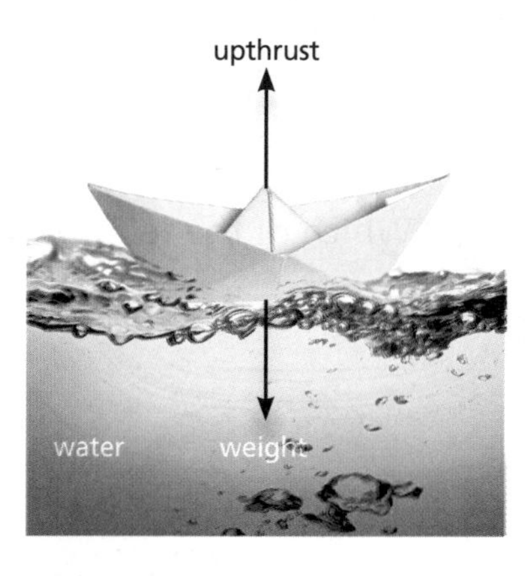

Floating and Sinking

- An object will float if the upthrust equals the weight of the object.
- If the weight is greater than the upthrust, the object will sink.
- Another important consideration is the **density** of the object. The greater the density, the more likely it will be that the weight will overcome the upthrust from the liquid and the object will sink.
- Ships float even though they may weigh thousands of kg because the weight is spread over a large area.

Upthrust / downthrust of blocks in water

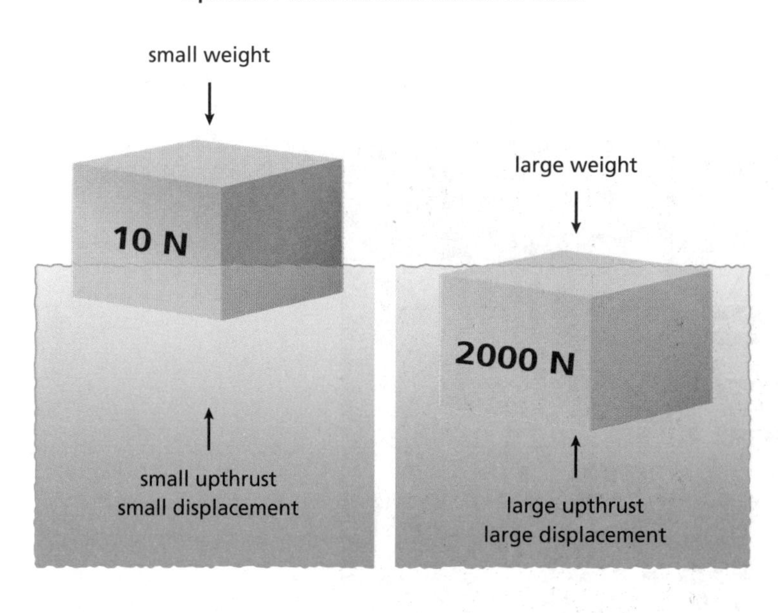

Key Point

Remember – when the forces acting on an object are balanced it will continue to do what it was doing. If it is floating, it will continue to float.

Quick Test

1. Why does atmospheric pressure decrease with height?
2. What is the unit for pressure?
3. If an object is floating with weight of 1000 N what is the upthrust?
4. For an object to float higher in water, what must be increased?

Key Words

pressure
upthrust
plimsoll line
density

Obtaining Useful Materials

1. Which of the following metals was discovered thousands of years ago? [1]

Gold

Aluminium

Iron

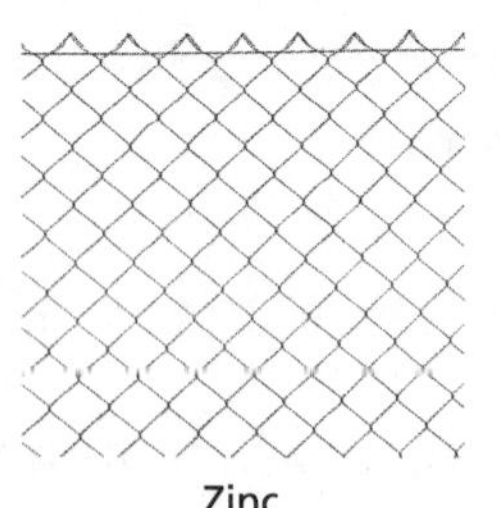
Zinc

2. a) Predict which **two** of the following reactions will take place.

 i) copper + aluminium nitrate **ii)** iron + copper sulfate

 iii) iron + potassium nitrate **iv)** copper + silver nitrate [2]

 b) For each of the reactions you selected, write the complete word equation for the reaction. [2]

3. The photo shows a blast furnace. A blast furnace is used to extract iron from its ore.

 Explain what happens when carbon and iron are reacted together in the blast furnace. [3]

Using our Earth Sustainably

1 Match the rock description to the type of rock:

Rock type	Rock Description
Igneous	has layers of crystals
Sedimentary	has veins, has a waxy appearance and is often used for statues
Metamorphic	is very hard and made of lots of small crystals

[3]

2 Copy the table and use the words below to complete the gaps, showing the abundance of elements in the lithosphere.

aluminium **iron** **oxygen** **silicon**

Element	Abundance (%)
	46
	28
	0.8
	0.5
Other elements	24.7

[4]

3 A group of friends are talking about recycling.

a) Which two friends are giving reasons **for** recyling? [1]

b) Write a counter argument to address Kalum's point. [2]

Forces and their Effects

1. Four Newton meters are shown in the diagram below. Which shows the highest amount of force? [1]

A
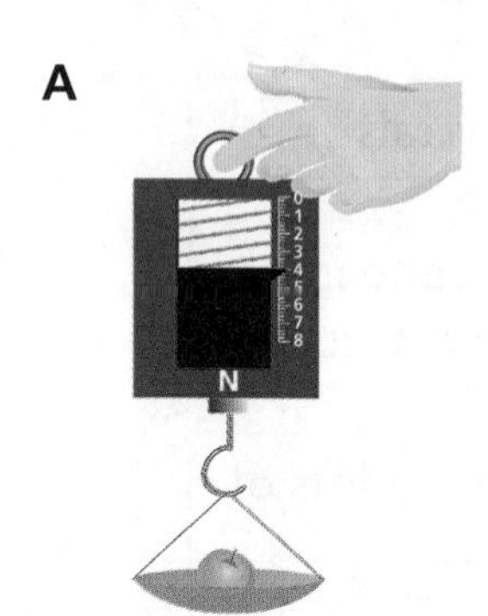

B

C
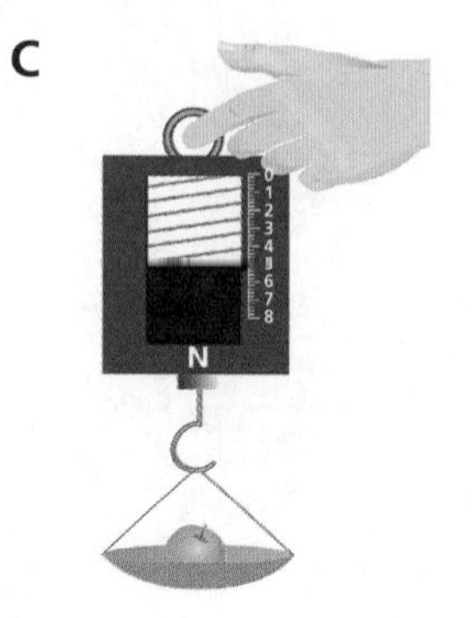

D
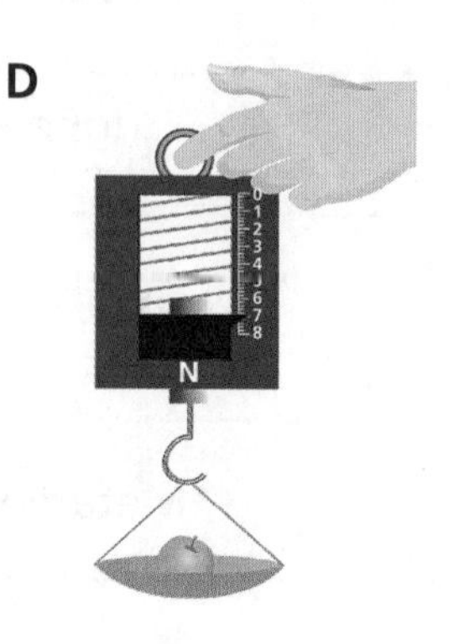

2. Sandy is playing with balancing scales. On the left-hand scale she has a ball and on the right she adds mass until it balances.

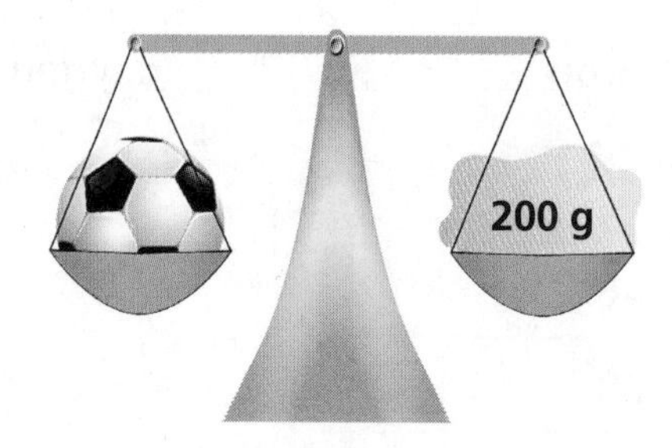

a) If she adds 200 g, what is the mass of the ball in grams? [1]

b) She now removes the ball and puts two blocks on the left-hand scale and 300 g on the right hand scale to balance it. What is the mass of one block in grams? [1]

3. Arran is training to run the 800 m in the Olympics. In her last race she won in 1 min 48 s. What was her average speed for the race? [2]

4. Joss is prising open the lid on a tin of paint. Calculate the **turning moment** for the force applied to the screwdriver.

Show your working and give the correct unit. [3]

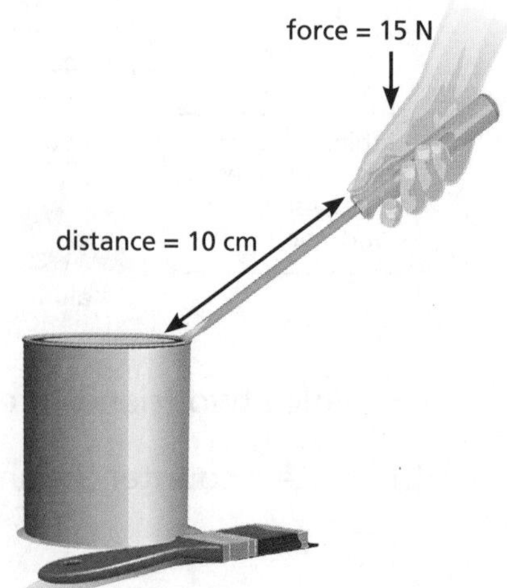

Exploring Contact and Non-Contact Forces

1 Fiesa brings two magnets together. They repel.

a) Copy and complete the diagram to show the poles on the magnets.

Magnet 1 Magnet 2

[2]

b) What would Fiesa have to do to make the magnets attract? [2]

2 Taking the gravitational field strength on Earth to be 10 N/kg, what would each of the following masses weigh?

10 kg 15.5 kg 2000 g [3]

3 The bottle of fizzy drink is full. The bottle has three holes made at points A, B and C.

a) Draw what would happen to the flow of fizzy drink at points A, B and C on the bottle. [1]

b) Explain why the flow at points A, B and C is different. [2]

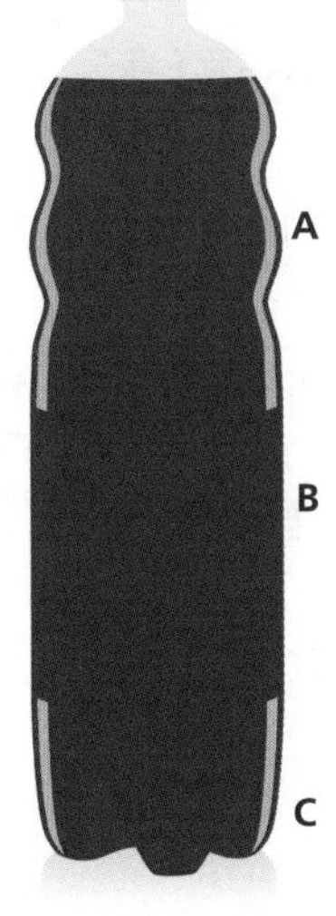

4 Amy and George are both the same weight, 100 N. Amy wears snowshoes with an area of 250 cm^2 and George wears skis with an area of 350 cm^2. Calculate the pressure exerted by each of them on the snow. [2]

Motion on Earth and in Space

You must be able to:

- Interpret distance–time graphs
- Explain and apply concepts of balanced forces and equilibria to analyse stationary objects
- Explain relative motion.

Describing Motion

- The motion of an object (the journey it takes) can be described by drawing a distance–time graph.
- The axes of the graph must be labelled correctly.
- Time is plotted on the *x*-axis.
- Distance is plotted on the *y*-axis.

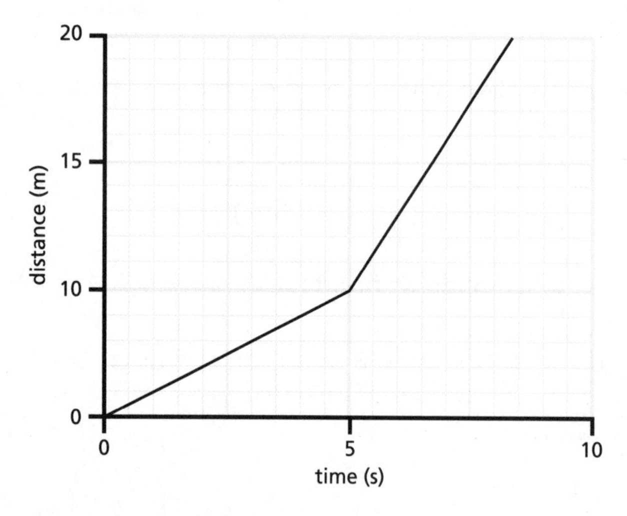

- The line of a distance–time graph represents the speed the object is travelling at:
  - A steeper line means more distance is covered in the same time, i.e the speed is faster
  - A shallower line means less distance is covered in the same time, i.e. the speed is slower
  - When the line is horizontal it means the object is not moving at all. It has stopped.
- A more complex distance–time graph appears below:

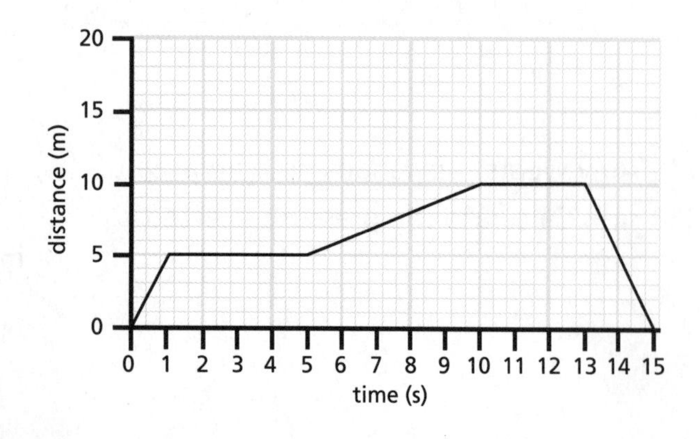

Key Point

The line of a distance–time graph shows the speed of an object.

Relative Motion

- The motion of objects is always **relative** to the **observer**.
- If two trains, A and B, are travelling at the same speed on tracks parallel to one another it would appear to an observer on either train that both trains were at a standstill.
- If the trains were travelling on parallel tracks towards each other at the same speed, then an observer on either of the trains would get the impression that the other train was travelling at twice the speed of their train.
- To calculate relative velocity to an observer:
  - If the object is moving towards the observer, add the speeds.
  - If the object is moving away from the observer, subtract the speeds.
- So for trains moving parallel to each other:

  Observer → 10 m/s
  Object → 10 m/s

  relative velocity = 10 m/s – 10 m/s = 0 m/s
- For trains moving in opposite directions:

  Observer → 10 m/s
  Object ← 10 m/s

  relative velocity = 10 m/s + 10 m/s = 20 m/s

Key Point

The position of an observer may influence what they see. Everything is relative!

Key Point

Velocity is the same as speed, but with direction added.

Forces in Equilibrium

- When forces act in opposite directions to each other and are the same size, they are balanced or in **equilibrium**.
- When the forces are balanced, the object will continue to do what it is doing:
  - If it is moving, it will move at a constant speed.
  - If it is stationary, it will stay stationary.
- A spring with a 10 N weight attached will stretch until the force of the spring pulling the weight upwards (the **reaction force**) equals 10 N.

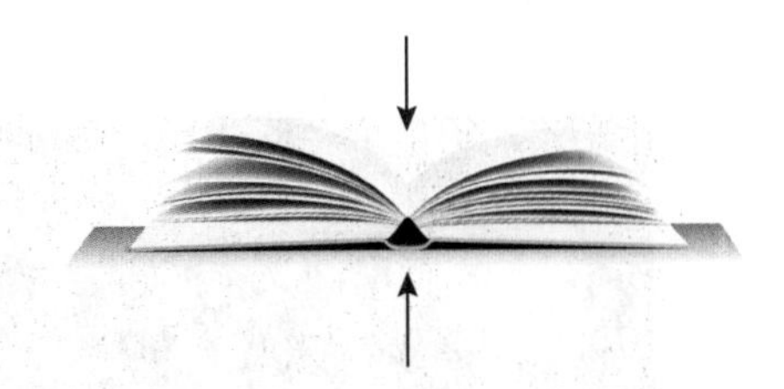

Quick Test

1. If an apple produces 1 N of force on a desk, it is not moving. What is the reaction force?
2. What can a distance–time graph tell us?
3. One car travels left at 70 mph and another travels right at 60 mph. What is the relative speed?
4. What does the slope tell you on a distance–time graph?

Key Words

relative
observer
equilibrium
reaction force

Physics

Motion on Earth and in Space

You must be able to:

- Explain differences in gravity on different planets
- Explain why the Earth has seasons
- Explain why day length varies on both the Earth's hemispheres.

Space and Gravity

- Gravity is a force that gives objects weight.
- Every object with mass has gravity.
- Gravity, however, is a relatively weak force and a very large mass is needed before the gravitational pull is noticed.
- Therefore, all planets and moons (not just Earth) have a gravitational pull.
- Weight is calculated by the formula:

weight = mass x gravitational field strength (*g*)

- The Earth has a **gravitational field strength** (*g*) of 10 Newtons per kilogram, which is written as: ***g* = 10 N/kg**
- On other planets and moons the value of *g* will differ.

Planets, showing approximate gravitational field strength values (N/kg)

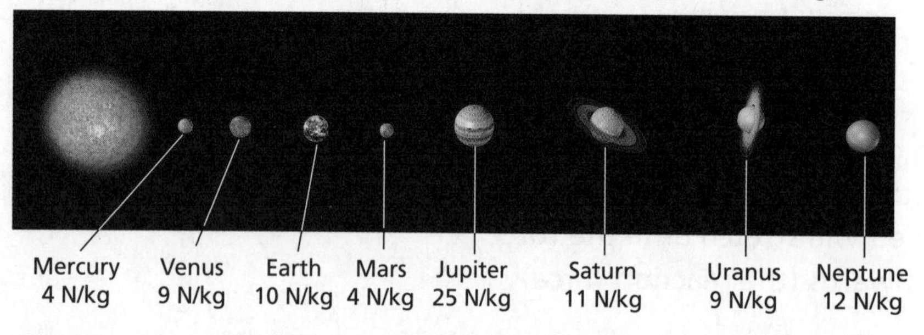

- The gravitational field strength of the Earth keeps the Moon in orbit, whilst the much larger gravitational field strength of the Sun keeps the Earth and all the other planets in orbit around it.
- The Sun is our closest star and it orbits the centre of our galaxy, the Milky Way.
- The Milky Way is filled with billions of other stars, all orbiting the galactic centre due to the gravitational pull from a black hole in the centre.
- Our galaxy is only one of billions, all exerting a gravitational pull on each other.
- Distances in space are enormous, so scientists use the astronomical unit for distance, the **light year**.
- A light year is the distance that light travels in 1 earth year, 9,460,730,472,580,800 km or 9.5 trillion km.

Key Point

Do not confuse mass and weight. Weight is a force and changes depending where you are. Mass always stays the same.

The Earth in Space

- The Earth spins on its axis once every 24 hours.
- The Earth is tilted on its axis, so at different times of the year the Northern and Southern Hemispheres receive different amounts of radiation from the Sun. This leads to seasons.
- Day length varies in the two hemispheres as the planet orbits the Sun.

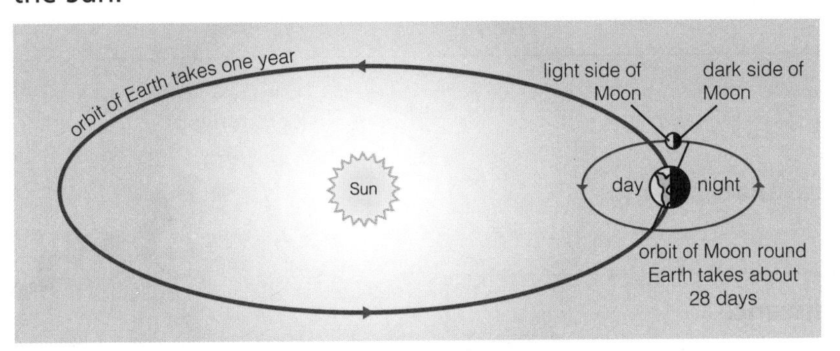

- An **equinox** is when both hemispheres of the Earth receive the same amount of light, so day and night are the same length wherever you are on the planet.
- There are two equinoxes a year, on the 21st March and 21st September.
- Between 21st March and 21st September, days are longer than nights in the Northern Hemisphere. In the Southern Hemisphere the opposite is true.
- After 21st September, day length shortens in the Northern Hemisphere and lengthens in the Southern Hemisphere.
- For the Northern Hemisphere 21st June is the longest day (summer **solstice**) and 21st December the shortest day (winter solstice). This is reversed for those in the Southern Hemisphere.

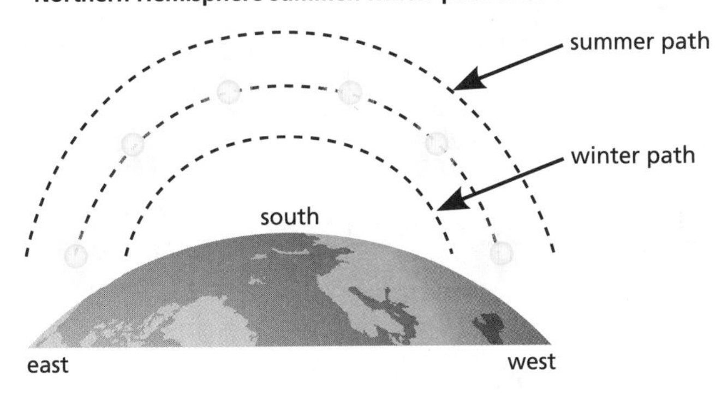

Quick Test

1. How long does daylight last on 21st March (the spring equinox) in the Northern Hemisphere?
2. Why do we get seasons?
3. If a person weighed 800 N on Earth, what would they weigh on Jupiter, where the gravitational field strength is 25 N/kg?
4. Suggest why g is only 4 N/kg on Mercury.

Key Point

If the Earth was not tilted then there would be no seasons. There would be no difference in day length across the planet.

Key Point

The dates given for the equinoxes and solstices are the average. The precise dates vary according to the year.

Key Words

gravitational field strength
light year
equinox
solstice

Energy Transfers and Sound

You must be able to:
- Understand and apply the model of energy transfer to various contexts
- Explain the role and significance of fuels.

Energy Changes Due to Forces

- When force is applied, work can be calculated using the formula:

work = force x distance

- Increasing the distance to the pivot means less force is applied, but over a larger distance. Reducing the distance increases the force used. The amount of work always remains the same.
- Similarly, gears increase the distance to the pivot, with a smaller gear turning many times, making a larger gear move.
- When an object changes its motion, other changes are experienced, e.g. dropping an object causes changes in the energy in the object. The energy is transformed from gravitational potential to kinetic energy.

GPE = 100% KE = 0%	GPE = 50% KE = 50%	GPE = 0% KE = 100%
1 The instant the object leaves the hand all of its energy is in the form of **gravitational potential energy**.	2 When the object is half-way to the ground, half the energy will be gravitational potential energy and the other half **kinetic energy**.	3 At the instant the object hits the ground, all of its energy is in the form of kinetic energy.

- A dynamo is a device that converts the movement of a wheel into electrical energy to power a lightbulb or recharge a battery cell by transforming kinetic to electrical energy.
- The more efficient a transfer, the less wasted energy there is.

Energy Changes Due to Altering Matter

- When forces are applied to an object they will change its shape.
- When **compression** is applied to a spring, energy is stored as elastic energy.
- When the spring is released, it returns to its original shape.

Key Point

Often some of the energy is 'lost' as heat and sound. We say it is transferred.

- The elastic energy is transformed to kinetic energy as it decompresses.
- Fuel has its energy stored in the form of **chemical potential energy**.
- When a source of ignition is provided the energy in the fuel is released through combustion and this happens very quickly:

fuel + oxygen → carbon dioxide + water + energy

- The energy released is heat, but some energy may be transformed into light.
- There are many types of fuels, storing different amounts of chemical potential energy.
- Most of the fuels in current use derive from crude oil although the number of alternatives, such as biodiesel, are increasing.
- The metabolism of food is similar to the combustion of fuel.
- The process inside cells is called respiration. It happens more slowly than combustion and involves food (the fuel) reacting with oxygen:

food + oxygen → carbon dioxide + water + energy

- When hot and cold objects come together, heat energy will transfer into the cooler object, until both objects are the same temperature.

Key Point

You can't let the cold in, only the heat out!

Energy Changes Due to Vibrations and Waves

- The Sun warms the Earth by radiation.
- Radiation is the transfer of heat energy via waves (infrared) that can travel in a vacuum.
- The objects do not need to be touching to receive the heat energy.
- **Vibration** of atoms occurs when objects get hot.
- The more kinetic energy, the more vibrations there are and the hotter they get.

Energy Changes Due to Electricity

- In an electric circuit the battery stores chemical potential energy.
- When the circuit is completed, the chemical potential energy is transformed into electrical energy – carried by the electric charge that moves around the circuit.

Quick Test

1. What is the word equation for the combustion of fuel in oxygen?
2. Describe the differences in reaction rate between combustion and respiration.
3. What is meant by radiation in terms of energy transfers?
4. An object has gravitational potential energy of 1000 J. What will its maximum kinetic energy be?

Key Words

gravitational potential energy
kinetic energy
compression
chemical potential energy
vibration

Energy Transfers and Sound

You must be able to:
- Explain how sound waves behave
- Explain how different animals hear
- Explain how sound is produced.

Sound Waves

- Sound waves carry energy through a medium.
- The medium must have particles to transfer energy as sound.
- The closer the particles, the faster the energy can be transferred.
- Sound travels fastest in solids, slower in liquids and slowest in gases.
- The speed of sound in air is 340 m/s, in water it is 1500 m/s and in solids such as wood, 4000 m/s.
- The **frequency** of a wave is measured in hertz (1 Hz = 1 wave per second).
- Musical notes have separate frequencies.
- Animals can hear a range of sounds.
- As humans get older, the range of hearing decreases, with the highest frequency sounds being lost first.
- Objects produce sound when they vibrate. For example:
  1. A guitar string when plucked causes a vibration in the wire.
  2. The wire hits air particles and causes them to move.
  3. The air particles collide with other air particles. Eventually they make the ear drum vibrate at the same frequency.
- Loudspeakers work in a similar way:
  1. Music is converted to an electrical signal which causes electromagnets to move a fabric skin or membrane.
  2. The fabric skin's motion causes air particles to move.
  3. Eventually the sound waves reach the ear.
- Microphones work in the opposite way:
  - Sound waves produced by an instrument or voice hit a membrane attached to magnets
  - The motion of the magnets causes a changing electrical signal that can be recorded and used to make a loudspeaker move.

Key Point

A vacuum does not contain particles, so sound cannot travel in a vacuum.

Hearing range of animals

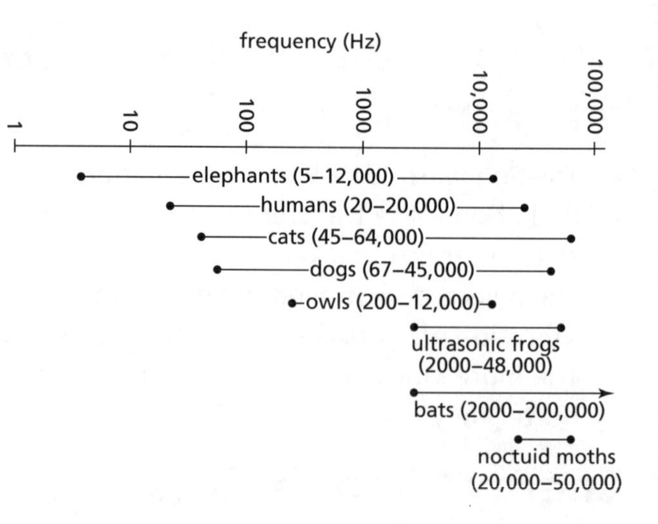

Sound from loudspeaker

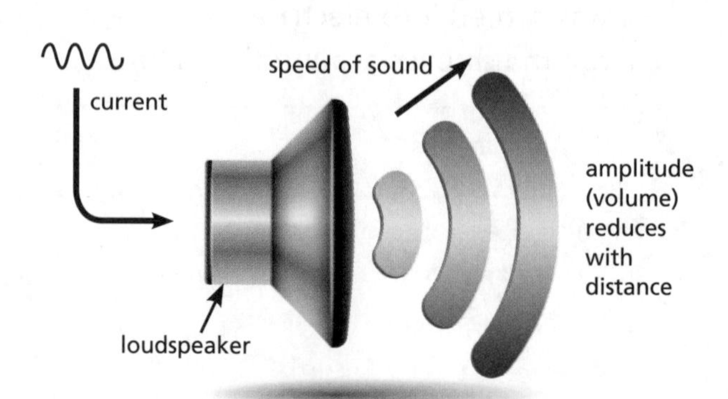

Using Sound

- Very high frequency sound (greater than 20000 Hz) can be used to clean objects, e.g. jewellers can clean jewellery using an ultrasonic bath. The high frequency vibrations cause minute particles of dirt to be displaced.
- **Ultrasound** is also used in medicine, e.g. to treat injuries and to detect the movement of a developing foetus in the mother's uterus.

Echoes and Absorption

- Sound waves behave like any other wave, which means that they can be reflected:

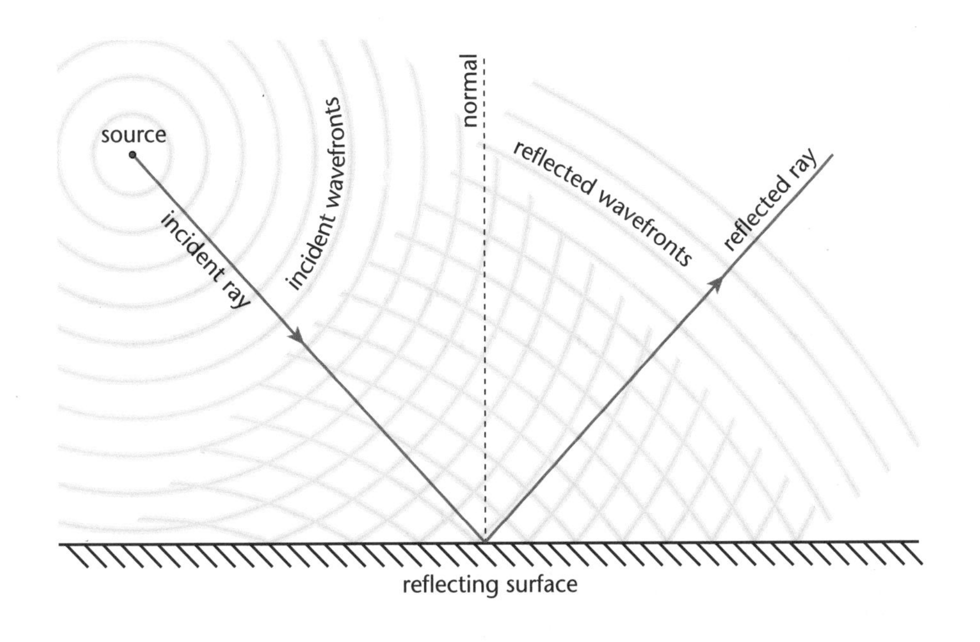

- When sound waves reflect straight back an **echo** is heard.
- When echoes would be a problem, e.g. in a music recording studio, steps are taken to prevent any echoes from spoiling the recording.
- Putting special angled tiles onto the walls and ceilings ensures that the sound waves keep reflecting within the tile until the energy has been **absorbed**.

Key Point

For an echo to be heard, the sound must reflect off an object. The smoother the object, the better.

Quick Test

1. What does it mean if a wave has a frequency of 50 Hz?
2. Explain how echoes occur.
3. Suggest how to reduce echoes in a sound studio.
4. Explain how a loudspeaker works.

Key Words

frequency
ultrasound
echo
absorb

Forces and their Effects

1. Look at the following diagrams showing an astronaut in space. For (a), (b), and (c), decide whether the astronaut will move or not. [3]

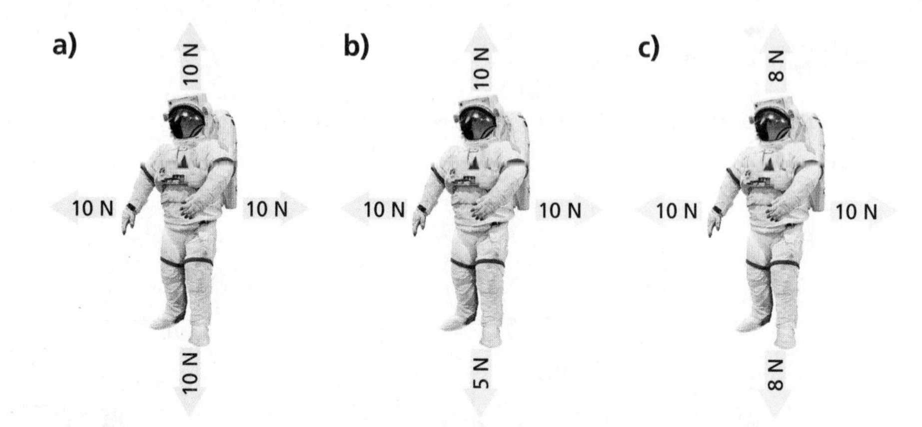

2. Ges is investigating whether a spring obeys Hooke's Law. He hangs masses from a spring and measures the length the spring stretches each time. The table below shows the results for Ges's experiment.

Mass (g)	Weight (N)	Stretch (cm)
0	0	0
100	1	2
200	2	4
300	3	6
400	4	8
500	5	10
600	6	
700	7	
800	8	16

Ges did not record a result for 600 g and 700 g. Predict what the stretch of the spring should have been (in cm) for these. [2]

3. The International Space Station (ISS) orbits the Earth at an average speed of 27,600 km/h.

Taking the distance of one orbit of the Earth as 42,927 km, how long does it take the ISS to orbit the Earth once? Show your working out. [3]

Exploring Contact and Non-Contact Forces

1. Which of the following forces are examples of non-contact forces?

   a) a tug of war | **b)** mass
   c) weight | **d)** static electricity [2]

2. Charlotte rubs a polythene rod with a cloth and holds it close to a stream of flowing water from the tap.

Normal flow of water

Flow changes with a charged rod

   a) Suggest why the water is attracted to the polythene rod. [2]

   b) Charlotte hangs the charged rod by a piece of string and then brings another rod with the same charge close to it. What will happen to the hanging rod? [1]

3. A heavy lifting ship sinks below sea level so that it can pick up large objects such as oil rigs. When it sinks, ballast tanks fill with water. The heavy lifting ship does not sink to the bottom of the sea.

   Explain, using the photo, why the ship does not sink. [3]

Motion on Earth and in Space

1. Lisa is walking her dog. The distance–time graph below shows Lisa's journey.

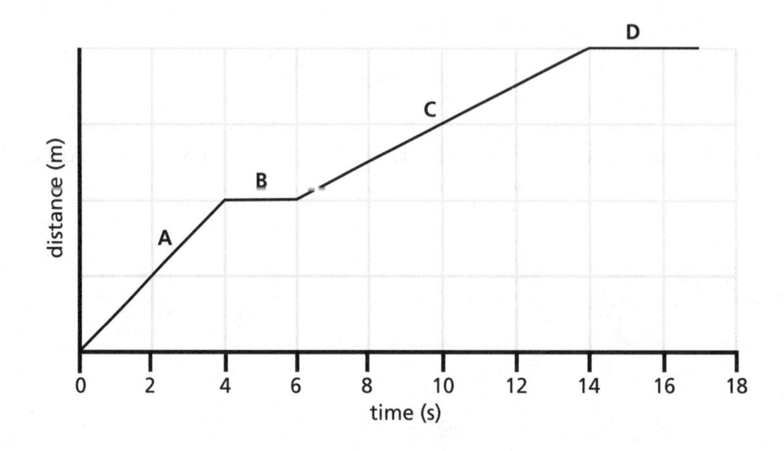

Each of the following describes part of Lisa's journey. Choose which of the labels A–D match with the parts of Lisa's journey given below:

Lisa stopped for 2 min **Lisa walked for 8 min**

Lisa jogged for 4 min **Lisa stopped for 3 min** [4]

2. Two cars are involved in a head-on collision. Luckily the occupants of both cars were not injured.

If both cars were travelling at 40 km/h, what would the relative speed have been? [1]

3. The closest star to Earth outside the solar system is Proxima Centauri, which is 4.2 light years from the Earth.

a) What is a light year? [1]

b) Taking the speed of light to be 300,000 km/s in a vacuum, calculate how far away Proxima Centauri is from the Earth in km. Show your working. [3]

4. The diagram below shows the path the Sun takes in the Northern Hemisphere in June.

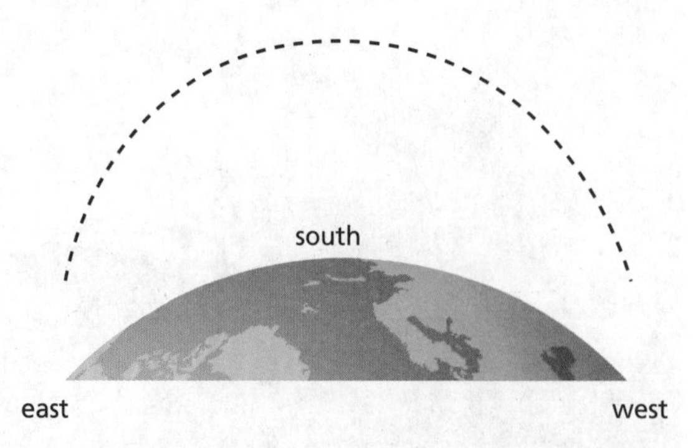

Copy the diagram and add the path that the Sun will take in December. [2]

Energy Transfers and Sound

1. A bungee jumper jumps off of a bridge.

   a) At what point does the bungee jumper have the most kinetic energy? [1]

   b) At what point does the bungee jumper have the most gravitational potential energy? [1]

2. The equation below is the word equation for the combustion of a fuel burned in excess oxygen.

   Fuel + A → B + C + energy

   a) What are A, B and C? [2]

   b) What is the name of the process in cells that is equivalent to combustion? [1]

3. Which of the following is the way that the Sun heats the Earth?

   a) conduction

   b) convection

   c) radiation

   d) nuclear [1]

4. Jamie is talking to his wife Linda. He is standing in the kitchen and Linda is in the next room.

   Explain why Jamie and Linda can hear each other talking, even though they are not in the same room. [2]

Magnetism and Electricity

You must be able to:

- Explain electric current as a flow of charge
- Apply concepts of potential difference and resistance
- Describe and measure current in series and parallel circuits
- Use formulae to calculate resistance, current, voltage and energy used.

Electric Current

- In an electric circuit, charged electrons move through the wire and components.
- The rate of the flow of charge is called the electric current (I).
- Current is measured in amperes (A) using an ammeter.
- The battery in a circuit provides energy to the charged electrons passing through.
- The battery cell has a negative and positive terminal.
- The potential difference (p.d.) is the work done to move a unit of charge from one point to another in a circuit.
- In an electric circuit diagram the wires and components are drawn in a simple way to show the connections and components.
- A **series** circuit has a single loop:

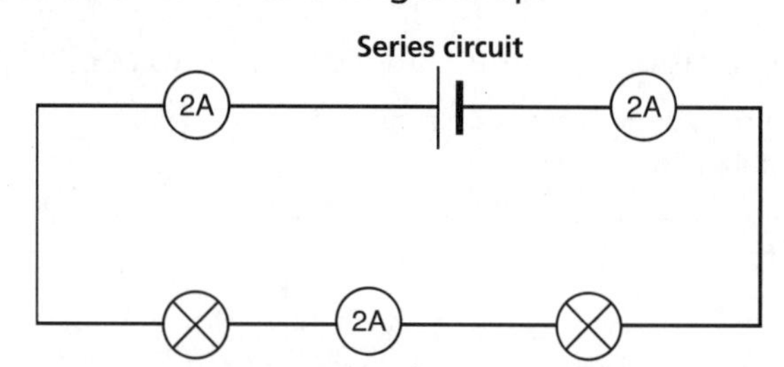

- In **parallel** circuits the flow of electric current is split between different branches. When the branches meet up, the currents add together again.

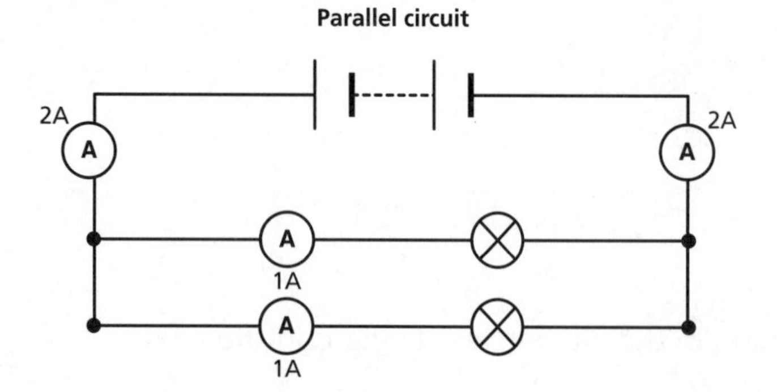

- Electricity in the home is connected as a parallel circuit, called a ring main.

Key Point

Potential difference is also known as voltage.

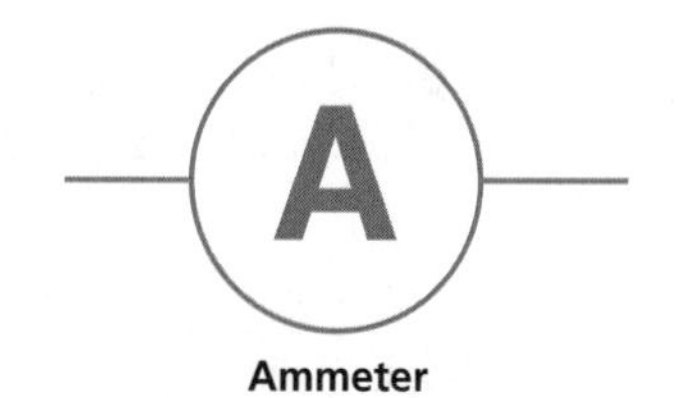

Ammeter

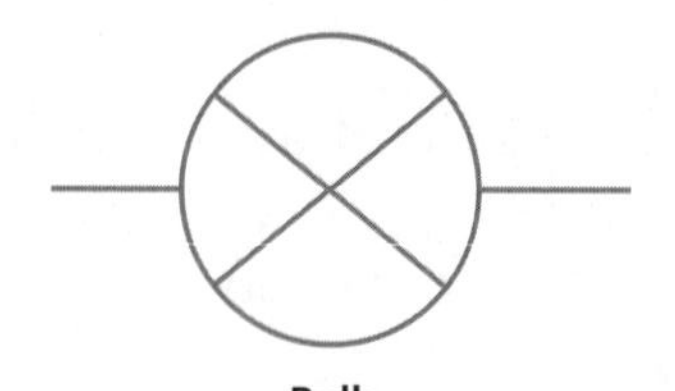

Bulb

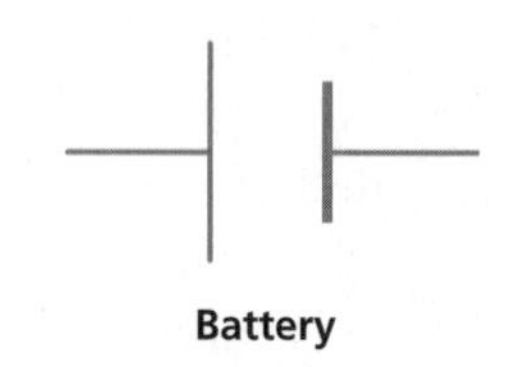

Battery

Components and Resistance

- All components in a circuit resist the flow of electric current; they have a **resistance**.
- An electrical **insulator** has complete resistance, allowing no current to flow through. A **conductor** has low resistance to the flow of current.
- Some components are engineered to be electrical insulators under certain conditions and conductors in others, e.g. light dependent resistors.
- Resistance is measured in ohms (Ω).
- The higher the resistance, the slower the flow of electric current through the component, and the energy carried by the charged electrons has time to transfer to the component.
- A light bulb has a resistance, so electric charge can pass energy to the filament of the bulb, which produces light as it heats up.

Key Point

Resistance is the ratio of the p.d. to the current.

Electricity Calculations

- Resistance (R) can be calculated using the following formula:

R = V / I

- Where R = resistance (Ω), V = potential difference (V) and I = current (A).
- The current in a series circuit is the same all the way around the circuit.
- Electric power (P) is measured in watts (W) and is calculated using the formula:

P = V x I

- Amount of energy transferred is calculated using the formula:

energy transferred (kWh) = power (kW) x time (h)

Key Point

Kilowatts (kW, 1000 W) are used rather than watts because otherwise the numbers involved get large and difficult to handle very quickly.

Power Ratings

- All electrical equipment has a power rating (in W or kW) which enables you to work out how much they use. The higher the power rating, the more electricity used.
- The time the equipment is switched on for is also important.
- A 2 kW heater switched on for 4 hours uses 2 kW x 4 h = 8 kWh
- Electricity bills charge for electricity based on how many kWh of electricity have been used.

Quick Test

1. What is the resistance if the p.d is 1.5 V and the current 3 A?
2. What is the unit for resistance?
3. Explain what an insulator is.
4. Draw a series and parallel circuit, each with 2 bulbs.

Key Words

series
parallel
resistance
insulator
conductor

Magnetism and Electricity

You must be able to:

- Describe magnetic attraction, repulsion and fields
- Explain the Earth's magnetic field and how it can be used for navigation
- Describe how electromagnets work.

Magnets

- Permanent magnets are made from magnetic metals and alloys.
- The three magnetic metals are iron, cobalt and nickel.
- Magnets have a North and South-seeking pole at each end.
- When two magnets are brought near one another they will either attract or repel each other.

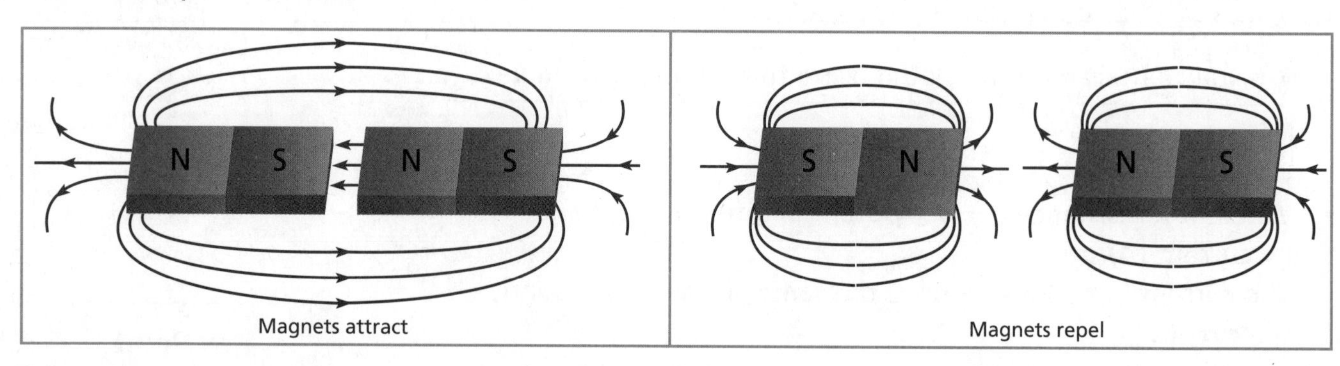

Magnets attract

Magnets repel

- The **magnetic field** of a magnet is invisible but can be shown using plotting compasses or iron filings sprinkled onto paper.
- The magnetic field can be represented by drawing **field lines**.

Key Point

The closer the field lines, the more powerful the magnetic field.

The Earth as a Magnet

- The Earth behaves like a bar magnet because it has a core made of iron and nickel.

Earth's magnetic field

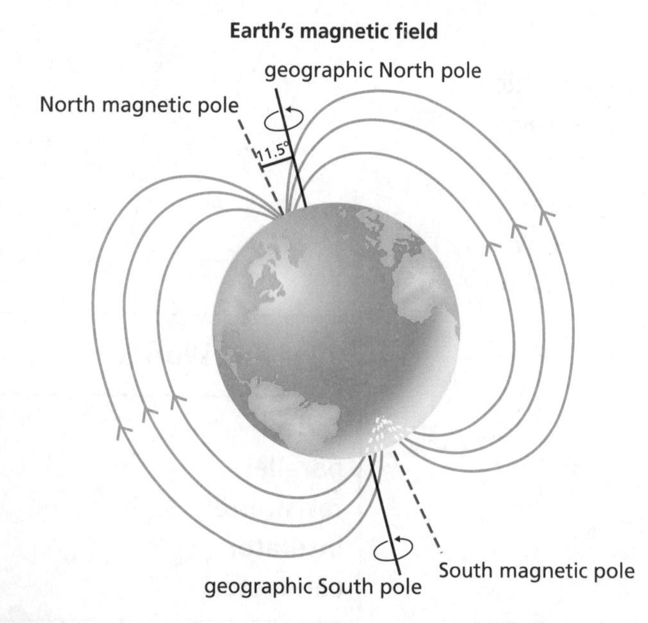

- The poles of the Earth are the equivalent of the poles of a bar magnet.
- Navigation is possible using a handheld compass, which seeks magnetic North.

Electromagnetism

- When an electric current is passed through a wire, it causes a weak magnetic field to be formed.
- The strength of the magnetic field can be increased by:
  - coiling the wire
  - increasing the current flowing through the circuit
  - adding a core made from a magnetic metal, e.g. iron.
- As the magnetic field is temporary and created by an electric current, the magnet formed is called an **electromagnet**.
- Electromagnets are stronger than permanent magnets and can be turned off.

Magnetic fields through an electric coil

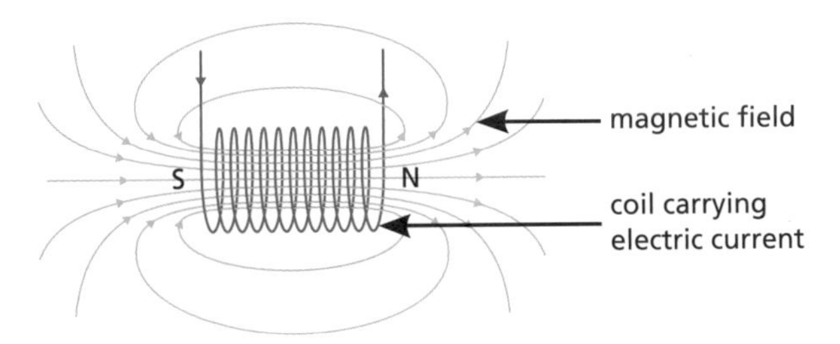

Key Point

You could test the strength of an electromagnet by seeing how many paper clips it could pick up.

Uses of Electromagnets

- Electromagnets are used in a number of different ways, e.g. heavy lifting in car breakers yards.
- In a **DC motor**, the wire and core can move freely.
- Brushes enable the wires to make contact without tangling.
- The motor will continue to turn for as long as there is a current.

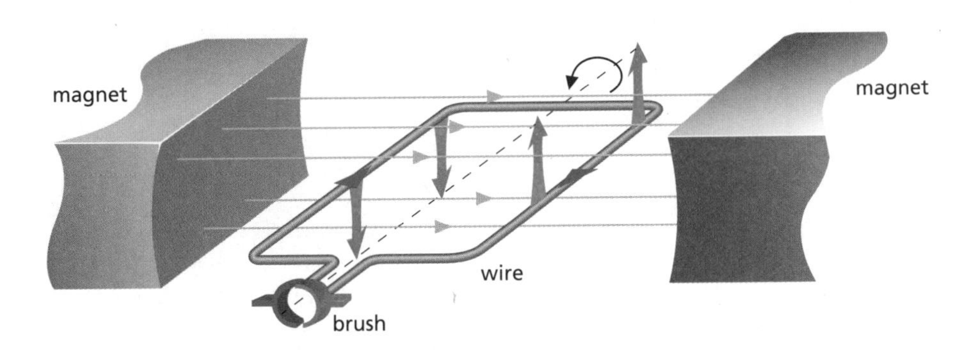

Quick Test

1. Name the ways an electromagnet's strength could be increased.
2. Suggest how the Earth's magnetic field is like a bar magnet.
3. What does it mean if the magnetic field lines are close together?
4. What are the three magnetic metals?

Key Words

magnetic field
field lines
electromagnet
DC motor

Physics

Waves and Energy Transfer

You must be able to:

- Explain how waves can be visualised using a ripple tank
- Explain how waves interact to produce interference
- Compare and contrast waves in sound and water with light waves.

Observing Waves

- It is possible to see the shapes of waves as they travel through water.
- The energy causes undulations which travel through the water.
- A ripple tank can be used to view the waves formed when a bar rapidly hits the water.
- When viewed from above the waves can be seen travelling with a **transverse** motion (at right angles to the direction of travel).
- The waves in water can also be reflected.

Key Point

Remember, the number of waves a second is the frequency. The more often waves appear, the higher the frequency.

Ripple tank with waves

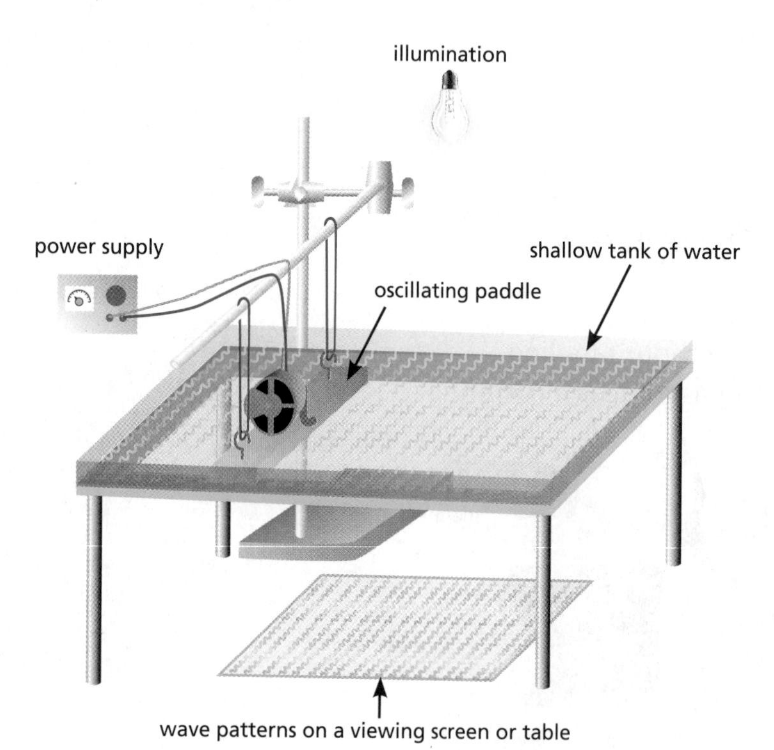

- A wave viewed from the side has the following features:
  - The height of the wave is the **amplitude**
  - The top of the wave is the **peak** .
  - The distance between a point on one wave and the same point on the next wave is the wavelength.

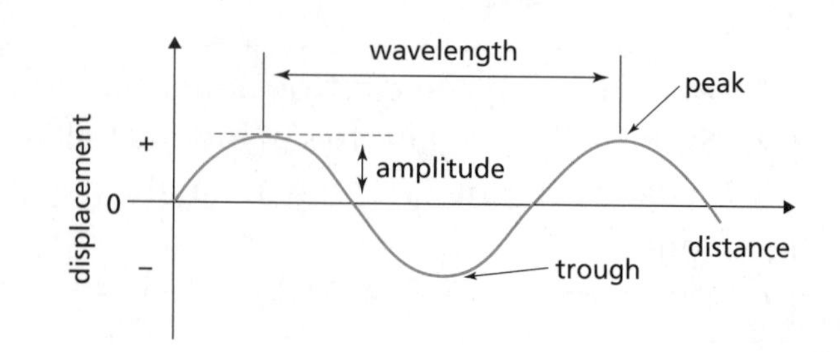

- Although the wave moves from left to right, a boat would move up and then down over the same spot.

Boat rotation on a wave

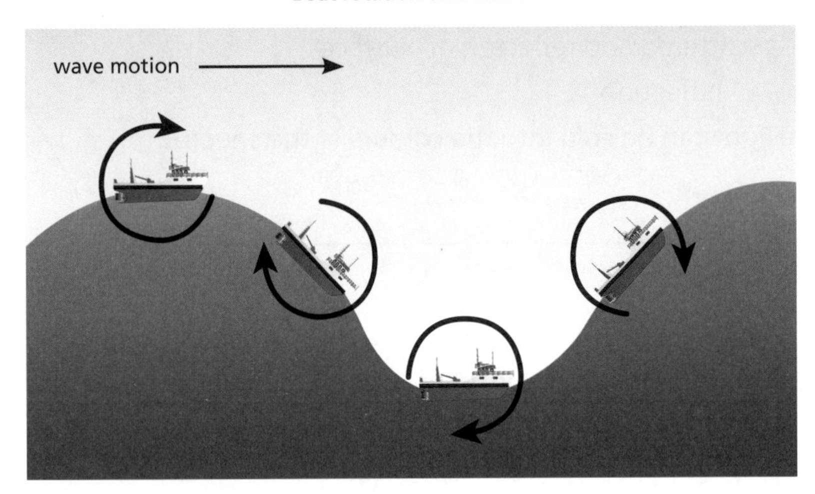

Superposition

- Waves can interact with each other. This is called **interference**.
- If the peaks and troughs arrive at exactly the same time as the other wave then they will combine to produce waves that are the sum of each contributing wave.
- If the waves arrive out of phase, then they will cancel each other out.

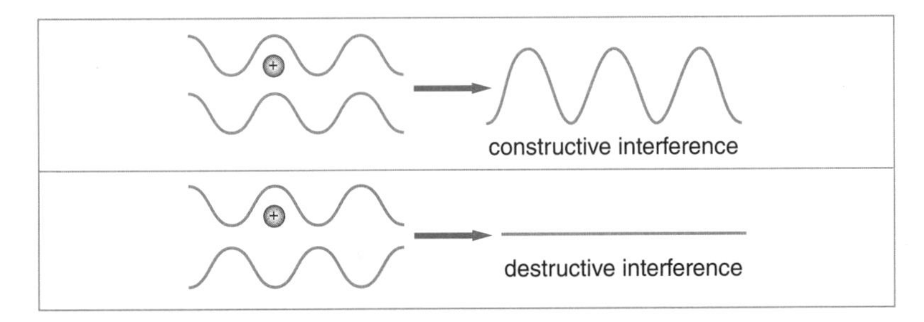

Key Point

All waves, including sound and light, can interact in this way.

Comparing Sound with Light

- Light travels in waves, but unlike sound or water waves, light does not need a medium (made of particles) to travel through.
- Light can travel in a vacuum (the absence of particles).
- The speed of light in a vacuum is 300,000 km/s.
- In other substances, such as air, water and plastic, the speed of light is slower.

Quick Test

1. What makes light waves different to sound or water waves?
2. What device can be used to look at waves in water?
3. Approximately how many times faster does light travel than the speed of sound in air?
4. What happens when two waves arrive in phase?

Key Words

transverse
amplitude
peak
interference

Physics

Waves and Energy Transfer

You must be able to:

- Explain how light is scattered and reflected off surfaces
- Understand the law of reflection
- Explain how white light can be split into the colours of the spectrum.

Light and Materials

- When light waves hit an object the light may be absorbed or reflected by the object.

Diffuse scattering and specular reflection

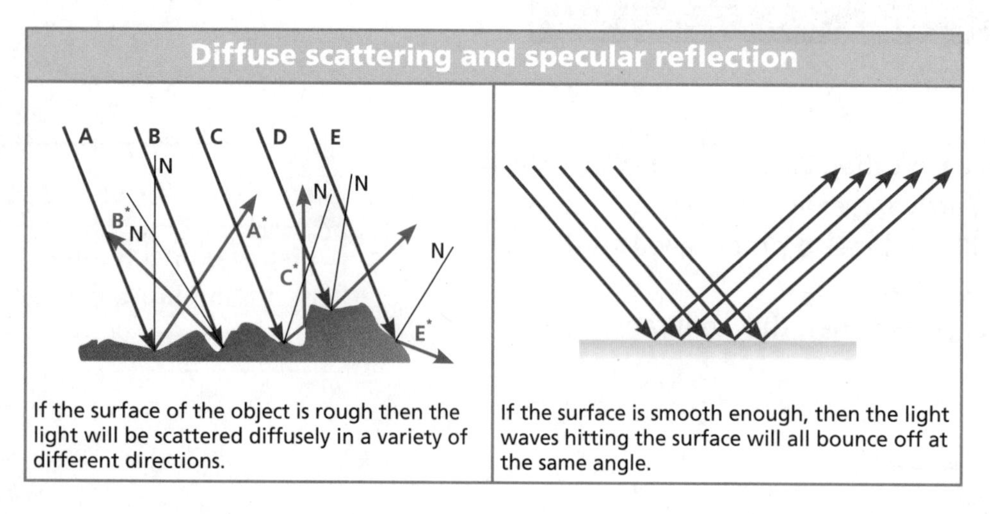

If the surface of the object is rough then the light will be scattered diffusely in a variety of different directions.

If the surface is smooth enough, then the light waves hitting the surface will all bounce off at the same angle.

Reflection

- Light waves can be drawn as rays.
- Light follows the law of **reflection**, which states that
  the angle of incidence = the angle of reflection
- An observer looking at the reflected rays will see an image.

Law of reflection | **Observer looking at reflected rays**

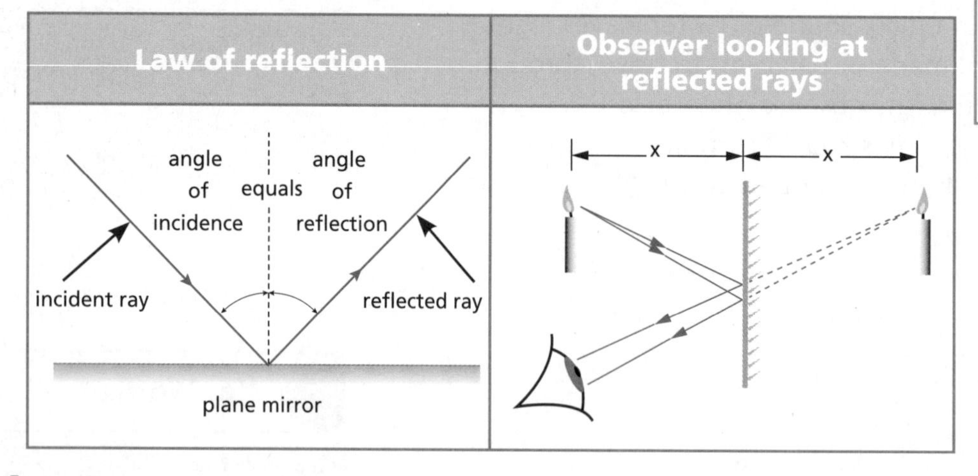

Key Point

Incident and reflected rays are compared to what a normal ray of light would do. A normal is drawn at right angles to the surface.

The Eye

- The most simple camera is called a pinhole camera.
- It has no lens, just a small hole for light to pass through.

- The eye acts like a pinhole camera.
- The image forms on the retina at the back of the eye. The image information is sent to the brain and the image reversed, so that it is 'seen' the right way up.

Eye as a pinhole camera

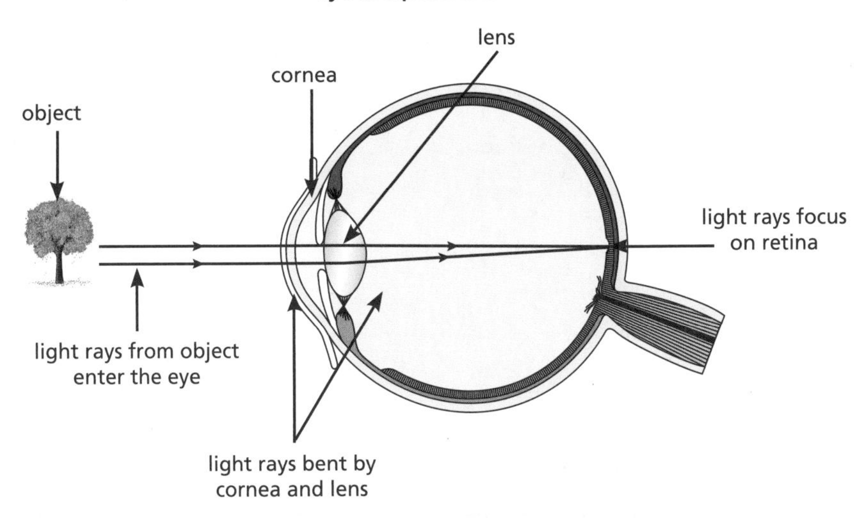

Focus of light on the retina

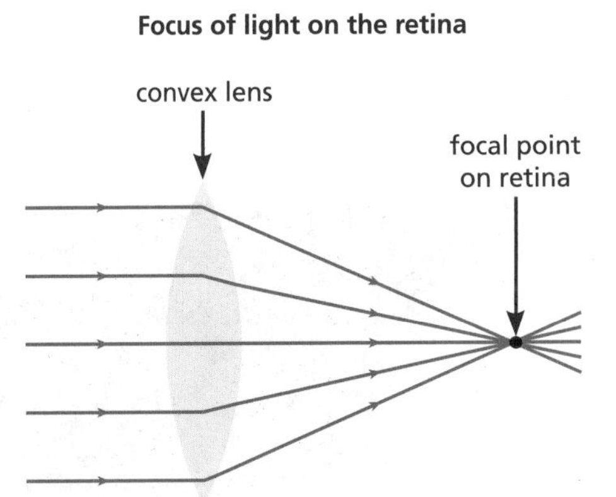

- A **convex** lens brings the light to a focus on the retina by a process called **refraction**.
- Refraction is where light slows as it passes through an optically dense medium or speeds up going into an optically less dense medium.
- Refraction causes the ray to bend.

Seeing Colour

- Light waves carry energy.
- Specialised cells in the eye are hit by the light ray, causing a chemical reaction. An electrical signal is then sent to the brain.
- With a camera the light hits a sensor which then sends an electrical signal to the memory card.
- Visible light is made up of waves of different frequencies.
- A prism splits white light into the colours of the spectrum, based on their **frequency**.
- Objects absorb different frequencies of light.
- The light that is not absorbed is reflected, giving the object colour.

Key Point

Although we can see millions of different colours, our eyes only have receptors for red, green and blue.

Quick Test

1. What colour is most refracted in white light?
2. What device can split white light into colours?
3. How does a pinhole camera work?
4. What is diffuse scattering?

Key Words

reflection
convex
refraction
frequency

Motion On Earth and in Space

1. Which of the following cars has balanced forces acting on it?

   a) accelerating to 80 km/h

   b) deccelerating to 20 km/h

   c) braking to avoid an accident

   d) travelling at 60 km/h [1]

2. The diagram below shows the planets in our solar system.

   a) The planets all orbit the Sun.

   What force keeps the planets in orbit? [1]

   b) It takes light 4.1 h to travel from the Sun to Neptune.

   Taking the speed of light to be 300,000 km/s, calculate the distance (in km) of Neptune from the Sun.

   Show your working out. [3]

3. Explain why planet Earth has seasons. [3]

Energy Transfers and Sound

1. An elephant is being lifted using a lever.

a) Ollie is trying to work out the best way to lift the elephant with the lever. Which of the following suggestions would enable him to lift the elephant more easily?

i) move the elephant further from the pivot
ii) use less force
iii) move the pivot closer to the elephant
iv) use more force [2]

b) When the elephant is lifted it gains energy. Which of the following types of energy will the elephant have after being lifted?

i) sound ii) mass iii) kinetic iv) gravitational potential [1]

2. In the UK an anti-loitering device was created to deter teenagers from hanging around shops in large groups. It emitted an extremely loud and annoying high pitched sound.

a) Suggest why teenagers could hear the anti-loitering device and why adults could not. [2]

b) The alarm used a loudspeaker that could vibrate at very high frequencies.
Explain how the sound travelled from the loudspeaker to a teenager's ear. [4]

3. Radar is a system that uses sound to reflect off objects. Radar is used to detect airplanes. The USAF Stealth bomber is an airplane that does not reflect sound.

Which of the following is the best explanation of why the Stealth bomber is invisible to radar.

a) The bomber reflects the radar signal the wrong way.

b) The bomber absorbs the radar signal.

c) The bomber is a good reflector of sound.

d) The radar travels through the bomber, so there are no reflections. [1]

Magnetism and Electricity

1. Draw a bar magnet, showing the magnetic field lines. [2]

2. Andy and Catherine are building a simple electromagnet.

   a) They want to increase the strength of the electromagnet. Which of the following would be ways to increase strength?

   i) more turns in the wire

   ii) use copper wire

   iii) use thicker wire

   iv) add an iron core

   v) add a plastic core

   vi) increase the current [3]

   b) Catherine wants to know why an electromagnet can be more useful than a permanent magnet. What should Andy tell Catherine? [1]

3. The parallel circuit below shows four ammeters. If the first ammeter (A1) reads 4 A, what would each of the other ammeters read? [2]

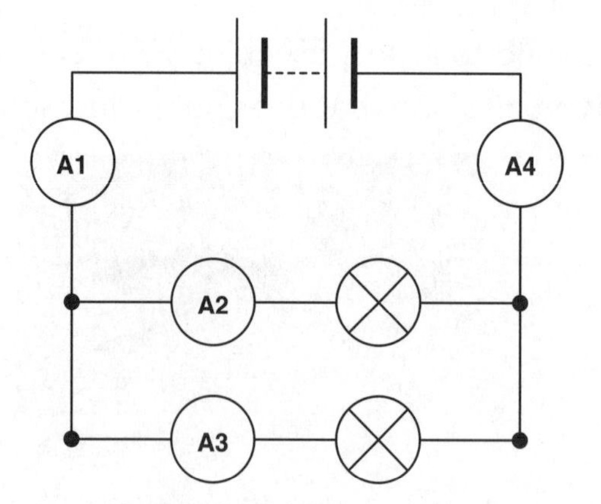

4. **a)** What current will flow through a 60 W bulb with a voltage of 230 V? [1]

   b) Calculate the energy (in kWh) transferred if the bulb was on for 200 hours.

   Show your working out. [3]

Waves and Energy Transfer

1 Naveen shines a laser at a mirror. She angles the laser so that it is at 30° to the normal.

Copy the diagram below and add the incident and reflected rays. [3]

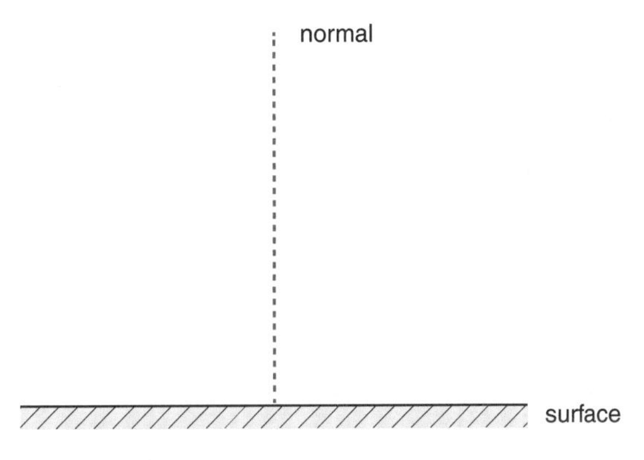

2 **a)** Copy the diagram below and add the labels **peak, trough** and **amplitude** to the waves shown. [3]

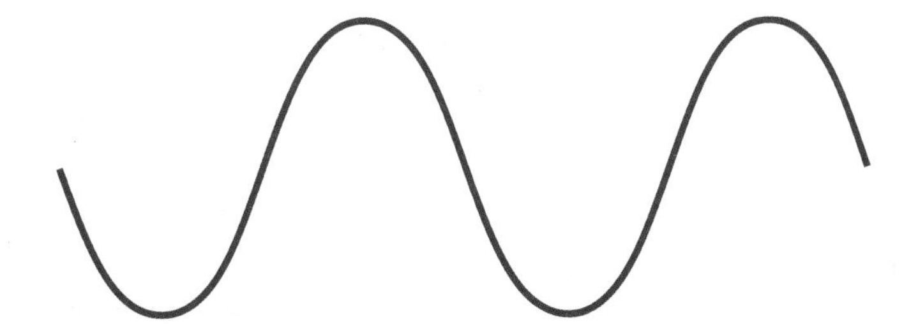

b) In the diagram, 1 second has passed. What will the **frequency** of the wave be?

Give the correct unit. [2]

3 White light is shone through a prism. What is the order of colours that appear on the screen? [4]

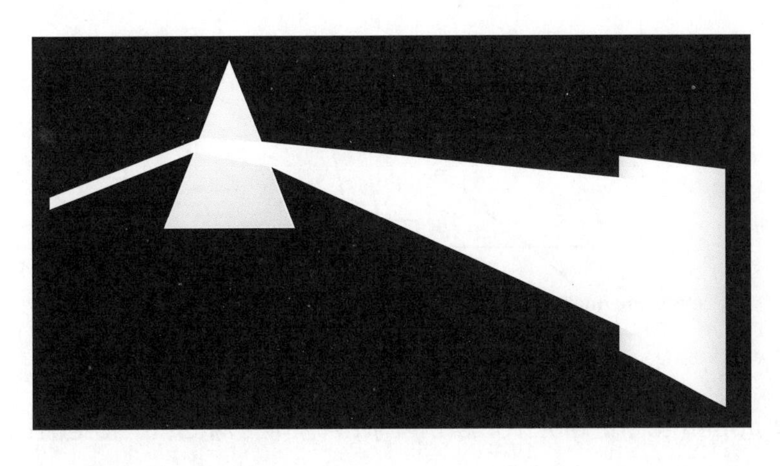

Magnetism and Electricity

1. Which of the following elements are magnetic?

   a) sulfur b) carbon c) cobalt d) vanadium [1]

2. Which of the following circuits will work? [1]

A
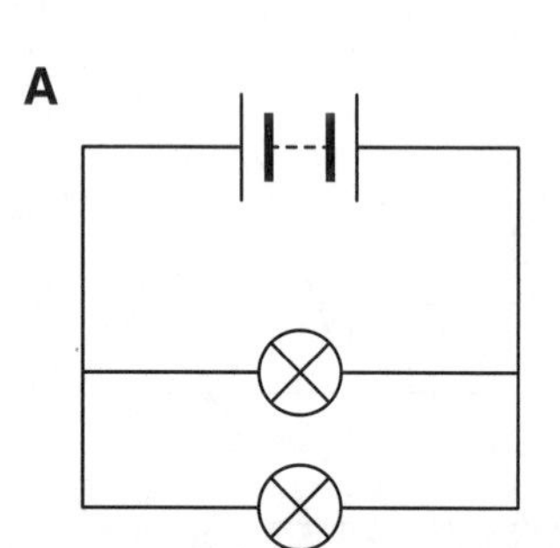

B
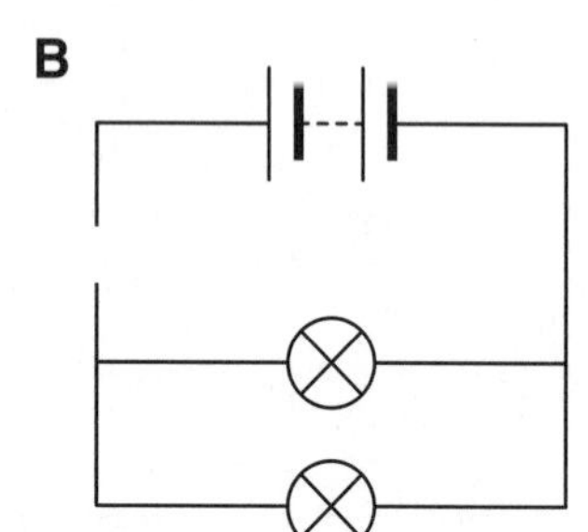

C
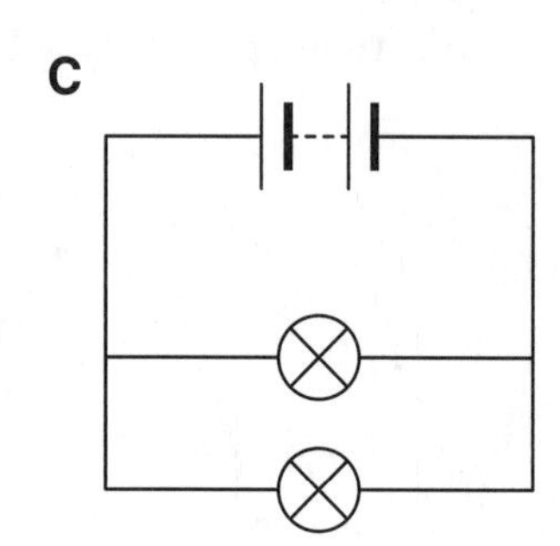

D
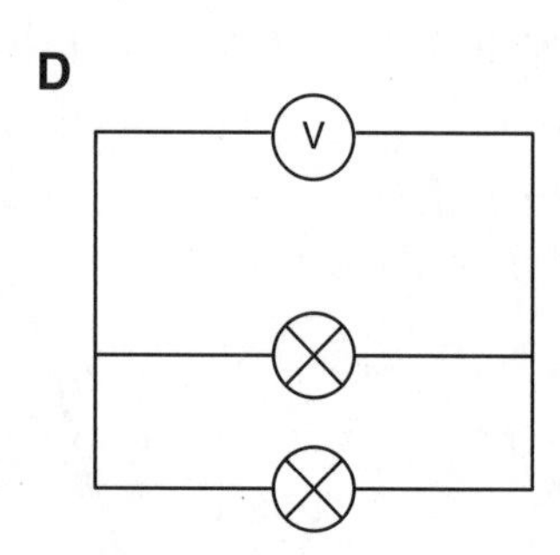

3. The diagram below shows a relay switch. It enables a small, safe circuit to be switched on and, in doing so, turn on a more dangerous circuit with much higher voltages.

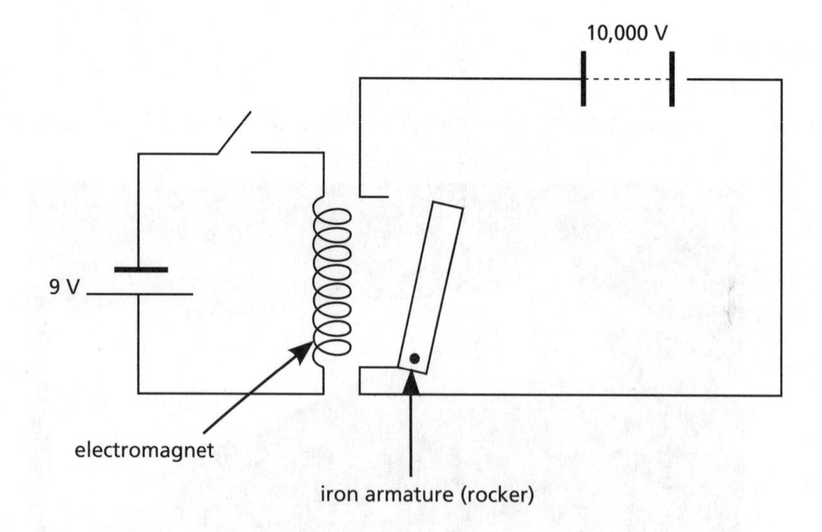

   a) What will happen to the electromagnet when the switch is closed? [1]

   b) Explain what will happen to the armature once the switch is closed. [2]

Waves and Energy Transfer

1. Explain why we can see our reflection in a mirror but not in a sheet of paper. [3]

2. Roger is using a ripple tank. He sets up the waves with a frequency of 10 Hz.

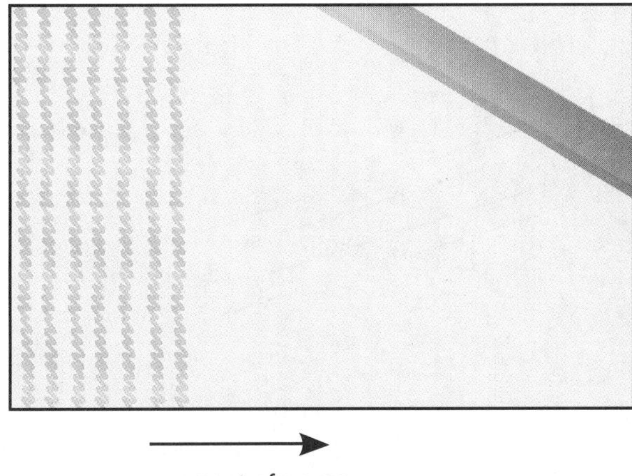

a) What does 10 Hz mean? [1]

b) Copy the picture and add four more waves on the ripple tank. [2]

c) Roger puts a barrier into the water at an angle.

On your diagram, draw the reflected waves. [2]

3. Sara is looking at a tree.

a) Draw the light rays from the top and bottom of the tree and show their route through the eye. [3]

b) What is the name of the process by which light is bent by the lens? [1]

Mixed Test-Style Questions

1 Quetzal is carrying out the distillation of alcohol. He sets up the apparatus as shown in the diagram below.

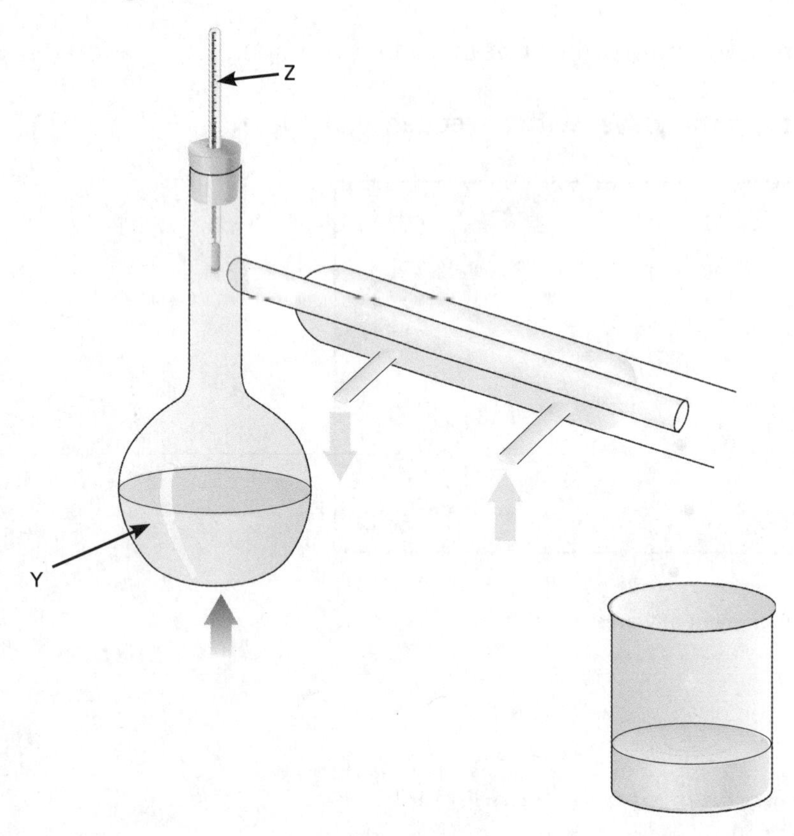

a) i) On the diagram above, write the letter B where the temperature will be the **highest** in the experiment.

1 mark

ii) On the diagram above, identify the distillate using the letter X.

1 mark

b) Suggest where cold water enters into the distillation tube and why it is important that the cold water flows in at that point.

....

....

....

....

....

2 marks

c) Ethanol is an alcohol.

It has the chemical formula C_2H_6O

How many **atoms** are present in a molecule of ethanol?

Tick the correct box.

2 ☐ 8 ☐

3 ☐ 9 ☐

1 mark

d) There are a number of state changes that take place during distillation.

In the boxes below draw the particle arrangement for the particles at point Y and at point Z.

Point Y

Point Z

1 mark

TOTAL

6

2 A group of performers are carrying out acrobatics for a TV talent show.

Michael and Kerrie lift a pole from each end and keep it level.

a) If Kerrie has to use 20 N of force to keep her end of the pole up, how much will Michael use to keep it level?

Tick the correct box.

80 N ☐

100 N ☐

20 N ☐

150 N ☐

1 mark

b) Amy vaults onto the middle of the pole and does a handstand.

Amy weighs 600 N.

How much force will Michael have to use to keep balancing the pole with Kerrie?

Show your working out.

....

....

2 marks

c) Nicola is wearing a pair of stilts.

Nicola has a weight of 650 N.

The end of each stilt has an area of 9 cm^{2}.

i) What **pressure** would the stilts exert on the ground?

Show your working out.

....

....

....

2 marks

ii) Describe what Nicola would need to do to exert **less** pressure on the ground.

....

1 mark

TOTAL

6

3 The lungs are where our body takes in oxygen from the air.

glass tube
glass jar
rubber tubes
balloons
rubber sheet

a) The diagram is a model that shows how the human lungs work.

Refer to the model to explain how it shows **breathing in** and **breathing out**.

Breathing in ....

....

....

....

....

3 marks

Breathing out ....

....

....

....

3 marks

b) The diagram shows the inside of three breathing airways.

One is **normal**, one is of a **heavy smoker**, and one is of someone with **asthma**.

Complete the label for each diagram.

Choose from these words: **normal** **smoker** **asthma**

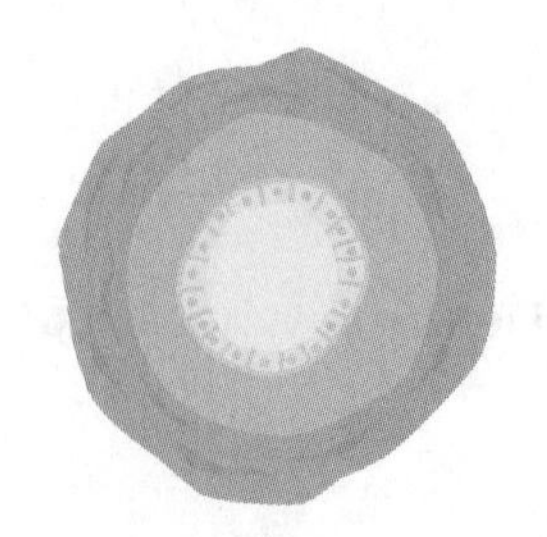

i) .... **ii)** .... **iii)** ....

3 marks

c) Oxygen moves from the lungs to the blood stream by diffusion.

The diagram shows the relative concentrations of oxygen in three different cells.

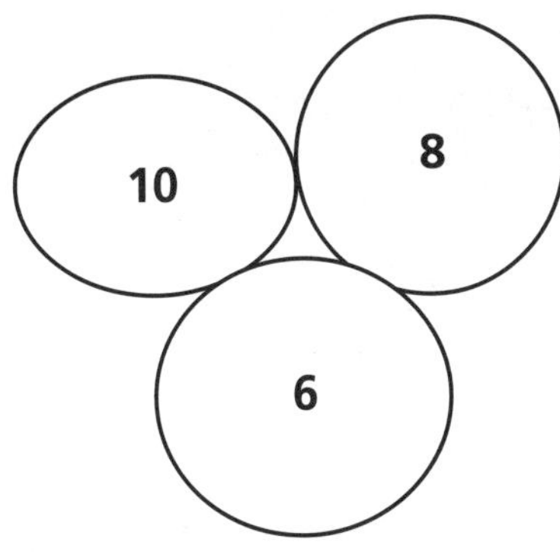

i) Draw three arrows (➔) on the diagram to show the movement of glucose between the three different cells.

3 marks

ii) Explain why oxygen moves between cells in this way.

....

....

3 marks

d) Oxygen is used by the body during aerobic respiration.

There are two types of respiration in humans; aerobic and anaerobic.

i) Other than the use of oxygen, describe two differences between aerobic and anaerobic respiration in humans.

....

....

....

2 marks

ii) Describe the difference between anaerobic respiration in humans and anaerobic respiration in yeast.

....

....

2 marks

TOTAL

19

4 Naveen is investigating how iron reacts with different chemicals.

In her first experiment she added an iron nail to different treatments to see what conditions are needed for rusting to take place.

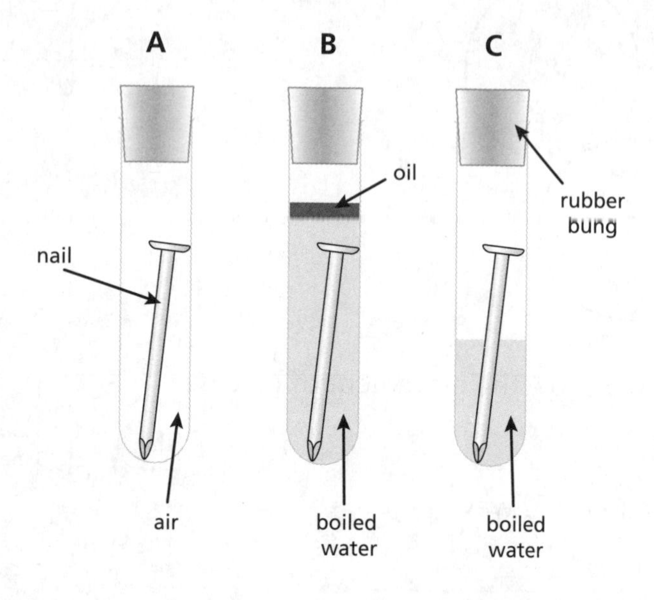

After a week she observed what had happened:

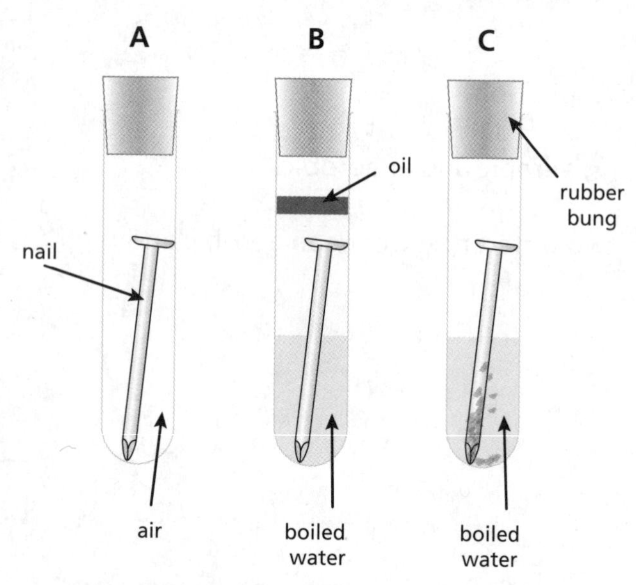

a) In one of the tubes a chemical reaction has taken place.

What evidence is there that a chemical change has taken place?

....

1 mark

b) Write the **word equation** for the rusting of iron.

.... + ... + ... → ....

2 marks

c) Suggest what Naveen's **hypothesis** would have been for this experiment.

2 marks

d) Iron is more reactive than copper.

Naveen placed an iron nail into a test tube containing blue copper sulfate solution.

i) Suggest what Naveen would **see** in this reaction.

2 marks

ii) Explain your answer.

2 marks

TOTAL

9

Mixed Test-Style Questions

5 This question is about plants and animals.

a) Look at the diagram of a food web.

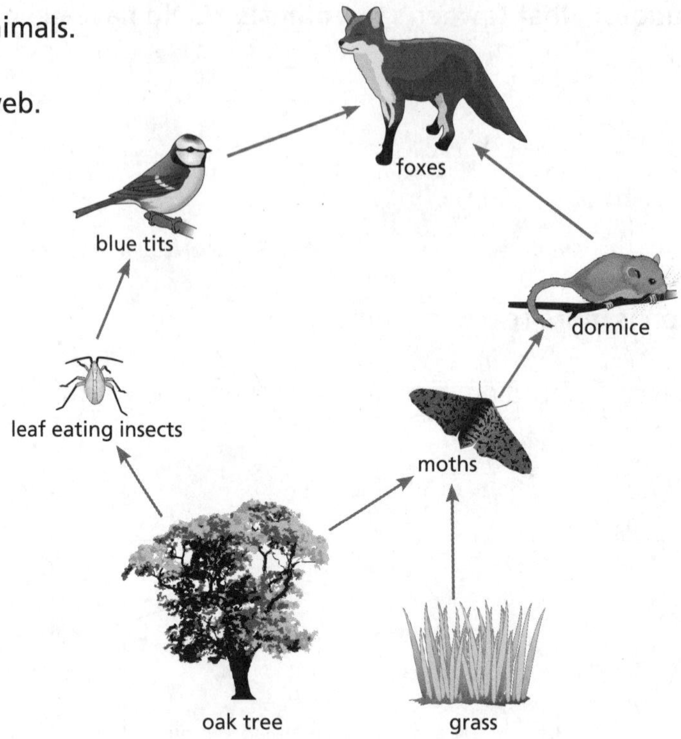

foxes

blue tits

dormice

leaf eating insects

moths

oak tree

grass

A farmer uses a pesticide to kill leaf-eating insects.

Suggest what effect a reduction of leaf-eating insects would have on the numbers of **blue tits** and **moths**. Explain your answer.

Blue tits ....

....

2 marks

Moths ....

....

2 marks

b) Both plants and animals can respire, but only plants can photosynthesise.

Read the statements about photosynthesis and respiration. Some are true. Some are false.

Put a tick (✓) in the boxes next to the true statements.
Put a cross (✗) in the boxes next to the false statements.

	✓ or ✗
Respiration stores energy as light	
Photosynthesis uses energy from light	
Respiration breaks down large molecules to smaller molecules	
Photosynthesis creates organic molecules from inorganic molecules	
Respiration releases energy from chlorophyll	
Photosynthesis stores energy as light	

6 marks

c) Many species of plants and animals are in danger of extinction.

i) Red kites are a type of bird found in England. Some people say that red kites were once extinct in England.

Explain why this is the wrong use of the word extinction.

....

....

1 mark

ii) Dinosaurs once existed. Now they are extinct.

Suggest what must happen to cause species such as dinosaurs to become extinct.

....

....

2 marks

d) Animals and plants are interdependent.

Complete the diagram to show how animals and plants depend upon one another for **oxygen** and **carbon dioxide**. Write the name of each gas in the correct space.

plants

______ ______

animals

1 mark

e) Many scientists are responsible for our understanding of the interdependence of plants and animals.

Which of these statements best describe how scientists work?

Put ticks (✓) in the boxes next to the **three** best answers.

Most scientific discoveries are the result of a scientist building on the work of a previous scientist	
Once a scientist has an idea they never change their mind	
Scientists never share their ideas with others	
Scientists have their results checked by other scientists	
Scientists use data from experiments to check their ideas	
Scientists never consider risk when doing experiments	

3 marks

TOTAL

17

6 The diagram below shows Katie's toy train.

The pieces are made of wood, but have a magnet at each end.

a) The magnet on carriage X **repelled** the magnet on carriage Y.

The magnet on carriage Y **attracted** the magnet on carriage Z.

i) **On the diagram above**, label the North and South poles of the magnets on X and Z.

1 mark

ii) Katie turned carriage Y around.

The carriage X and carriage Z were **not** turned around.

Describe what would happen now when Katie pushed the parts of the train together.

Explain your answer.

....

....

....

2 marks

Answers

Pages 4–11 Revise Questions

Page 5 Quick Test
1. cell wall, vacuole, or chloroplast
2. mitochondria
3. diffusion
4. cell, tissue, organ, system, organism

Page 7 Quick Test
1. sperm and egg
2. day 14
3. Insect pollinated — flower is brightly coloured, produces less pollen, produces nectar to attract insects. Accept reverse argument for wind pollinated flower.
4. animal, wind or self-dispersal

Page 9 Quick Test
1. The ribs move up and outward while diaphragm moves downwards.
2. volume decreases, pressure increases
3. diffusion
4. Smoking stops the mechanism for getting rid of mucus. Mucus builds up and causes coughing. Lung infections are more likely and there is a long term risk of cancer.

Page 11 Quick Test
1. Protein, fat, carbohydrate, vitamins, minerals, fibre, water.
2. Obesity, starvation, Kwashiorkor, vitamin deficiency, mineral deficiency, e.g. anaemia, vitamin deficiency, e.g. scurvy.
3. Mouth, oesophagus, stomach, small intestine, large intestine, rectum, anus.
4. Plants make their own food (photosynthesis), animals consume food.

Pages 12–13 Review Questions

Page 12

1. a) nutrition; growth; reproduction **[3]**

MRS GREN - This mnemonic can help you remember the characteristics of living things (Movement, Respiration, Sensitivity, Growth, Reproduction, Excretion, Nutrition).

2. b) i) **J** on shoulder, elbow, hip or knee **[1]**
   ii) **P** on skull or ribs **[1]**

Page 13

2. **a)** i; **b)** ii; **c)** iii; **d)** ii; **e)** ii or iii **[5]**

One material may have several different properties.

3. **a)** iii **[1]**;
   b) i; iii **[1 mark for both correct]**
   c) the Earth orbits the Sun – 365 days; the Earth rotates once – 1 day; the Moon orbits the Earth – 28 days **[3]**

Draw straight lines using a ruler.

Pages 14–15 Practice Questions

Page 14

1.

Membrane	Controls what enters and leaves a cell
Cytoplasm	Where chemical reactions take place
Nucleus	Stores information and controls the cell
Mitochondria	Releases energy from glucose
Cell wall	Supports the cell
Vacuole	Inflates the cell
Chloroplast	Changes light energy into food energy

[1 mark each up to a maximum of 7 marks]

Draw straight lines using a ruler.

2.

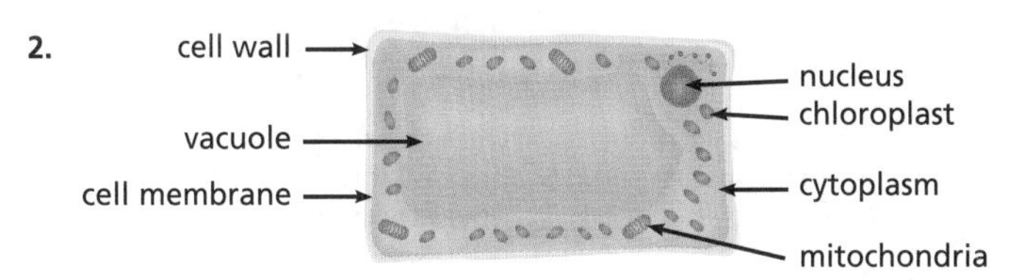

[1 mark for each label, up to a maximum of 6 marks]

3. iii **[1]**

Page 15

1. **a)** A diet that contains the right balance **[1]** and right amounts **[1]** of all the nutrients needed.
   b) Three of the following:
   eating too much – obesity
   eating too little – starvation
   not eating enough protein – kwashiorkor
   not eating enough vitamins – lack of different vitamins causes different diseases. Lack of vitamin C causes scurvy.
   not eating enough minerals – lack of iron causes anaemia. Lack of calcium causes soft bones. **[3 marks. 1 for each example and explanation]**

There is 1 mark for each example with explanation.

2. a)

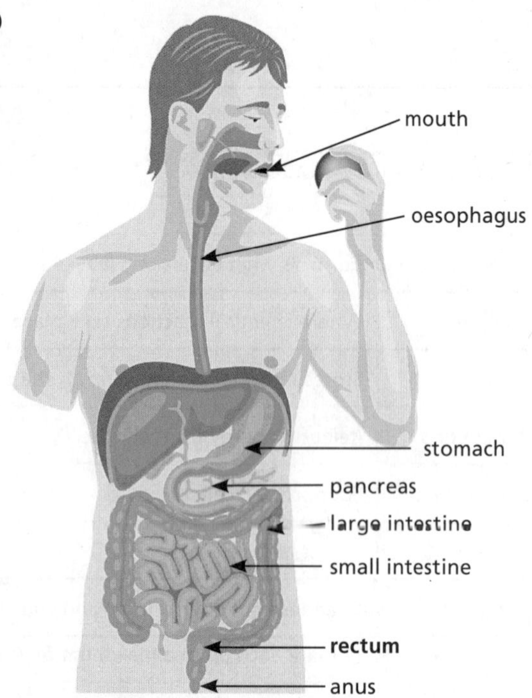

[8]

b) a) mouth – food broken into pieces
b) oesophagus – food passes from mouth to stomach
c) stomach – food mixed with acid and enzymes to help break it down
d) pancreas – produces enzymes to break down food
e) large intestine – water absorbed into blood stream
f) small intestine – broken down food absorbed into blood stream
g) rectum – waste material stored
h) anus – waste material eliminated [8]

3. feeding in animals involves eating food **[1]**; that is broken down **[1]** feeding in plants involves making food **[1]**; from simple substances, (by photosynthesis) **[2]**

State clearly whether you are writing about plants or animals.

Pages 16–23 **Revise Questions**

Page 17 Quick Test

1. aerobic
2. aerobic
3. anaerobic
4. anaerobic
5. alcohol

Page 19 Quick Test

1. support, protection, enables us to move, makes red blood cells
2. ligaments
3. tendons
4. Working against each other or in different directions.

Page 21 Quick Test

1. water + carbon dioxide $\xrightarrow[\text{chlorophyll}]{\text{light}}$ glucose + oxygen
2. Respiration uses glucose and oxygen to release energy and water and carbon dioxide are produced. In photosynthesis, carbon dioxide and water are converted into glucose and water using light energy.
3. have a large surface area to catch as much sunlight as possible; are green because of the chemical chlorophyll used in photosynthesis; have small holes called stomata on the underside of the leaf to let in carbon dioxide and let out oxygen; have tiny tubes called xylem to carry water and minerals up from the roots; have tiny tubes called phloem to carry glucose away for storage.

4. Plants make food and oxygen for animals. Animals make carbon dioxide for plants.
5. How different organisms rely upon each other for their survival.

Page 23 Quick Test

1. How energy moves through the food web.
2. Build-up of poisons towards the top of a food chain.
3. Variation is the differences between organisms. It increases the chances of survival when the environment changes.

Pages 24–25 **Review Questions**

Page 24

1. unicellular organisms need different structures to enable them to survive **[1]**; in different environments **[1]**

Look for key words. The key word in this question is 'explain'.

2. bone cell – skeletal system [1]
red blood cell – transport system [1]
nerve cell – nervous system [1]
sperm cell – reproductive system [1]

Draw straight lines using a ruler.

3.

Part	Male	Female
Testis	✓	✗
Egg cell	✗	✓
Vagina	✗	✓
Sperm	✓	✗
Penis	✓	✗

[5]

4. insect pollinated flowers are brightly coloured **[1]**, have a scent **[1]** and produce nectar to attract insects **[1]**, ***or*** wind pollinated flowers are not designed to attract insects so are not brightly coloured **[1]**, do not have a scent **[1]** or produce nectar but do produce lots of pollen **[1]**.
[total of 3 marks]

Show clearly if your answer relates to insect or wind pollinated flowers.

5. They pollinate our crops **[1]** and without pollination there would be no crops to harvest **[1]**.
6. Place pollen grains on a microscope slide. [1]
Cover pollen grains with a cover slip. [1]
Place slide on the microscope stage. [1]
Switch on lamp or adjust mirror. [1]
Select lens for suitable magnification. [1]
Focus the image with the focusing knob. [1]

Page 25

1. When we breathe in the ribs move up and out and the diaphragm moves down **[1]**. Pressure drops **[1]** and volume increases **[1]**. When we breathe out the ribs in and down and diaphragm moves up **[1]**. Pressure increases **[1]** and volume decreases **[1]**.
2. exercise; smoking; asthma [3]
3. carbohydrates – provide energy [1]
fat – stores energy [1]
proteins – used to grow new cells [1]
vitamins – needed for chemical reactions to take place [1]
minerals – needed for strong bones and blood [1]
fibre – speeds movement up through the gut [1]
water – dissolves chemicals for chemical reactions to take place [1]
4. Bacteria help break down food **[1]**, so it can be absorbed into the blood **[1]**.
5. It makes a reaction take place more quickly **[1]**, without being used up in the process **[1]**.

Learn definitions. They are an easy way to score marks.

6. The small intestine is where water is absorbed. **[1]**

Pages 26–27 **Practice Questions**

Page 26

1. uses oxygen – aerobic **[1]**
   produces lactic acid – anaerobic **[1]**
   produces alcohol – anaerobic **[1]**
   releases the most energy – aerobic **[1]**
   fermentation – anaerobic **[1]**

> Only write one tick in each row or you will not be given the marks.

2. In yeast it produces carbon dioxide **[1]**, and alcohol **[1]**. In humans it produces lactic acid **[1]**.
3. The skeleton provides support for the body **[1]**, enables movement to occur because of the joints **[1]**, protects the brain heart and lungs **[1]**, and makes red blood cells in the long bones **[1]**.
4. 1 – skull; 2 – ribs; 3 – spine; 4 – elbow joint; 5 – hip joint; 6 – femur; 7 – kneecap; 8 – radius and ulna; 9 – pelvis; 10 – humerus **[10]**
5. Muscles can only contract **[1]**, so two muscles are needed to move the joint in two different directions **[1]**. This is called an antagonistic pair **[1]**.

> Remember, muscles can only contract, they cannot expand.

Page 27

1. produces oxygen – photosynthesis **[1]**
   produces carbon dioxide – respiration **[1]**
   uses energy from sunlight – photosynthesis **[1]**
   releases energy – respiration **[1]**
   requires chlorophyll – photosynthesis **[1]**
2. any 5 of: very light; contains chlorophyll; has a large surface area; has holes to absorb carbon dioxide and release oxygen; has cells to carry water to the leaf and food away from the leaf **[5]**
3.

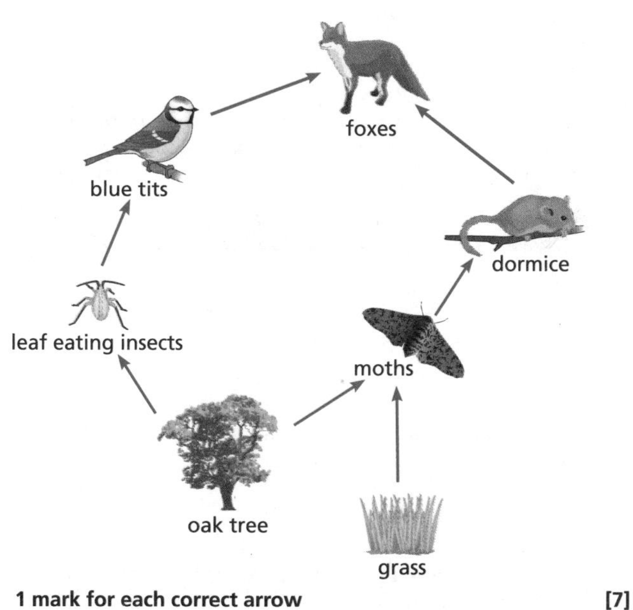

1 mark for each correct arrow **[7]**

> Always start with the plants and draw the arrows pointing away from them.

4. pollinate flowers **[1]**; may be a pest that destroy crops **[1]**.
5. Bioaccumulation **[1]** occurs because large animals eat lots of small ones so the poison builds up **[1]**.

Pages 28–35 **Revise Questions**

Page 29 Quick Test

1. 23 from mum and 23 from dad
2. chromosome
3. A single instruction found on a chromosome.
4. James Watson, Francis Crick, Maurice Wilkins, Rosalind Franklin

Page 31 Quick Test

1. Continuous, e.g. height, where a range exists between very short and very tall with most being somewhere near the middle of the range. Discontinuous, e.g. blood groups, where distinct groups exist.
2. sexual reproduction
3. The range of different organisms in an ecosystem.
4. A place where genetic material is collected and stored.

Page 33 Quick Test

1. A substance that affects the human body.
2. An unwanted effect on the human body.
3. Any two from caffeine, nicotine and alcohol.
4. The need to keep taking the drug.

Page 35 Quick Test

1. any three from: eyes produce tears that contain a chemical to kill microbes; ears produce wax to trap microbes; nose and throat produce mucus to trap microbes; skin acts as a barrier to microbes; stomach produces acid to kill microbes in food; urine flushes out microbes that enter the genitals; vagina produces acid to kill microbes
2. bacteria; viruses; fungi
3. when dead microbes are injected into the body causing the blood to make memory cells
4. antibiotics do not work against viruses

Pages 36–37 **Review Questions**

Page 36

1. Aerobic respiration uses oxygen **[1]**. Anaerobic respiration produces lactic acid **[1]**, produces alcohol **[1]**, releases the least energy **[1]**.
2. **a)** glucose; carbon dioxide **[2]**
   b) carbon dioxide; alcohol (can be either way round) **[2]**
   c) lactic acid **[1]**
3. Respiration releases energy **[1]** for all the chemical processes living things need **[1]**.

Page 37

4. supports the body; helps with movement; protects some organs; makes red blood cells. **[4]**

> This is a two mark question so you need to write two parts to your answer.

5. 1 = cartilage; 2 = lubricating fluid; 3 = ligament; 4 = tibia; 5 = femur **[5]**
6. **a)** contracts; bends arm **[2]**
   b) contracts; straightens arm **[2]**
   c) antagonistic **[1]**

1. **a)** carbon dioxide; oxygen **[2]**
   b) light; chlorophyll **[2]**
2. animals make carbon dioxide; plants use carbon dioxide; animals use oxygen; plants make oxygen **[4]**

Pages 38–39 Practice Questions

Page 38

1. 46; 46; 23; 23; 46 **[5]**
2. A – cell, B – nucleus, C – chromosome, D – gene **[4]**

Page 39

3. Continuous variation shows a complete range of differences **[1]** but discontinuous variation has discrete groups of variation **[1]**.
1.

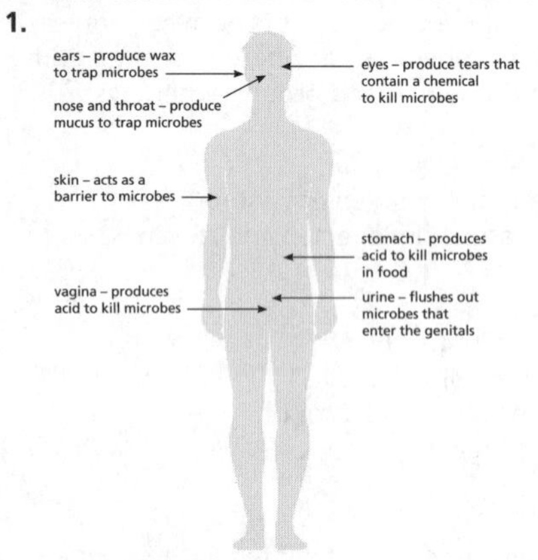

[7]

2. A medical drug is used to treat disease **[1]**. A recreational drug is used for pleasure **[1]**.
3. A depressant relaxes the body **[1]**, e.g. cannabis or heroin **[1]**. A stimulant makes you alert **[1]**, e.g. cocaine or amphetamine **[1]**. A hallucinogen alters the perception of reality **[1]**, e.g. LSD or magic mushrooms **[1]**.
4. Drinking can cause liver failure **[1]** and increased risk of heart failure **[1]**. Smoking increases the risk of heart disease **[1]** lung cancer **[1]** and lung infections **[1]**.

Pages 40–47 Revise Questions

Page 41 Quick Test

1. Filter the mixture using filter paper. The residue in the filter paper is the solid.
2. small
3. The evaporated and re-condensed liquid following boiling a mixture.
4. Draw a line approximately 1 cm from the bottom of chromatography paper in pencil. Place a sample of each liquid mixture on the line. Place the paper into a beaker containing the solvent (e.g. water). As the solvent moves up the paper the more soluble pigments move further. Once the first pigment has reached the top, remove and dry the paper.

Page 43 Quick Test

1.

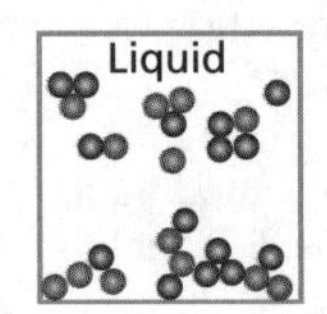

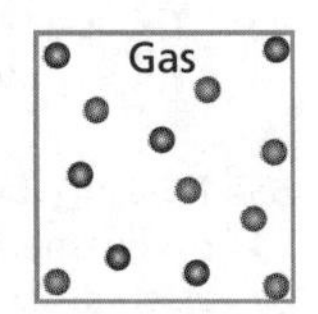

Gas

2. Calcium carbonate decomposes into calcium oxide and carbon dioxide when it is heated.
3. Oxidation is when substances gain oxygen in a reaction.
4. Water and oxygen. The process is faster if salt is also present.

Page 45 Quick Test

1. CO_2 and H_2O
2. The number of protons in the nucleus of an atom of the element.
3. two
4. three

Page 47 Quick Test

1. nitrogen + oxygen → nitrogen oxide
2. potassium sulfate
3. Three from: sonorous, shiny, malleable, conducts heat, conducts electricity, ductile, high melting point.
4. Three from: insulator (heat and electric), low melting points, gases at room temperature, low density.

Pages 48–49 Review Questions

Page 48

1. a) discontinuous **[1]**
   b) blood groups, tongue rolling **[2]**

Your teacher may give you other examples.

2. when the environment is changing very quickly **[1]**
3. a) Watson and Crick produced a theory to explain the structure of DNA **[1]**
   b) Rosalind Franklin produced evidence to support their theory **[1]**

Page 49

1. pathogen – disease-causing microbe **[1]**
   toxin – poison produced by a disease-causing microbe **[1]**
   antibody – chemical produced by white blood cells to kill pathogens **[1]**
2. bacteria – tuberculosis **[1]**
   virus – polio **[1]**
   fungus – athlete's foot **[1]**
   [1 mark for each named microbe with a disease caused by it]

Your teacher may give you other examples.

3. less addictive **[1]**
   less serious side effects **[1]**
4. Addiction is the need to take more and more of a drug **[1]**. Withdrawal is the physical effects on the body of not taking the drug **[1]**.

Learn definitions. They are an easy way to score marks.

5. Drinking alcohol – liver failure **[1]**
   Smoking – lung cancer **[1]**
   Taking LSD – trying to fly off a tall building **[1]**
   Using heroin – reduces breathing **[1]**
6. some white cells can engulf microbes **[1]**; some white cells produce chemicals to destroy microbes **[1]**; some white cells called memory cells protect us from future infections **[1]**

Pages 50–51 Practice Questions

Page 50

1. a) distillation **[1]**
   b) chromatography **[1]**
   c) filtration **[1]**
2. a) A and C **[2]**

Remember that elements that are gases at room temperature and pressure travel round as pairs.

   b) C **[1]**
3. Pour the sea water mixture into a filter funnel **[1]**. Allow the liquid to pass through **[1]**. Discard the residue and collect the filtrate **[1]**. Place the filtrate into an evaporating basin and heat **[1]**. Once the water has evaporated salt crystals will be left behind **[1]**.

The key here is to say what will happen in steps. A little like writing a recipe.

4. Vinnie could measure the distance each pigment moved on the chromatogram **[1]** from the starting line **[1]**.
5. The filter removes solid particles/waste/food that are building up in the water **[1]**. Filtration works by passing a liquid containing solid particles through a barrier that only allows small molecules through **[1]**, leaving the larger particles behind **[1]**.

Page 51

1. a) iii [1]
   b) ii [1]
2. ii; iii [2]
3. a) B [1]
   b) C [1]
   c) A [1]
4. An element is made up of only one kind of atom **[1]**.
   A compound is made up of more than one different type of atom **[1]** chemically joined together **[1]**.

Pages 52–59 **Revise Questions**

Page 53 Quick Test

1. In most solids the particles are closer together than in the liquid, therefore the density is greater.
2. Absolute zero (0 K) or –273 °C.
3. Sublimation is where a substance changes from a solid to a gas with no liquid phase.
4. Solids have all their particles touching and vibrating in position and have a defined shape. Liquids have at least 50% particles touching and move as well as vibrate and they take the shape of the container that they are in. Gases have no particles touching and will occupy all of the available space in a container.

Page 55 Quick Test

1. If the air in a sealed balloon is heated the gas particles will gain kinetic energy and move faster. This increases the air pressure, so the balloon will get bigger until the air pressure inside equals the air pressure outside. If the pressure becomes too great, the balloon may burst.
2. If the air in a sealed balloon is cooled the gas particles will lose kinetic energy and move more slowly. This decreases air pressure, so the balloon will get smaller until the pressure inside equals the air pressure outside.
3. Unlike other solids, the particles in ice are further apart than in a liquid. This means that there are fewer ice molecules in a given volume compared to the same volume of liquid water. The density of ice is therefore lighter, so it will float.
4. Diffusion occurs where particles at a higher concentration move randomly to areas where they are present in a lower concentration. The bigger the difference, the faster the rate of diffusion.

Page 57 Quick Test

1. A catalyst reduces the activation energy for a reaction, making it take place at lower energy levels than would normally be the case.
2. Activation energy is the minimum amount of energy needed to ensure that a reaction takes place.
3. A word equation just tells you what the reactants and products are, whereas a chemical equation tells you what atoms are present and the ratio of reactants and products.
4. (s), (l), (g), (aq)

Page 59 Quick Test

1. The substance is a weak acid, closer to being neutral than a strong acid.
2. Neutralisation is where an acid reacts with a base to form a pH 7 solution.
3. metal + acid → salt + hydrogen
4. UI paper relies on using your eyes and judging the pH, a pH probe and a data logger measures the actual pH. So pH probe and a data logger is both more accurate and more reliable.

Pages 60–61 **Review Questions**

Page 60

1. a) sample A = 3; sample B = 2 [2]
   b) sample A [1]
   c) The line is drawn in pencil because otherwise the line would move up the paper **[1]** along with the pigments being tested **[1]**.
2. a) thermal decomposition [1]
   b) calcium carbonate → calcium oxide + carbon dioxide [2]
3. $C(s) + O_2(g) \rightarrow CO_2(g)$ [3]

   [3 marks. 1 mark for correct reactants on left-hand side, 1 mark for correct product on right-hand side, and 1 mark for correct state symbols.]

Page 61

1. a) The circles (atoms) have been rearranged **[1]** and joined together in a new way **[1]**.
   b) Substance Y is oxygen/O_2 **[1]**. Substance Z is water/dihydrogen oxide/H_2O **[1]**.
   c) Mass has been conserved in the reaction because the number of atoms of each element is the same on both sides of the equation. [1]
2. A = sulfuric acid; B = magnesium nitrate; C = hydrochloric acid; D = hydrogen; E = lithium [4]
3. a) P;
   b) C;
   c) P;
   d) P;
   e) C [1]
4. $2Mg(s) + O_2(g) \rightarrow 2MgO(s)$ [1]

> Although d) is balanced, it is incorrect as oxygen exists as O_2 molecules.

Pages 62–63 **Practice Questions**

Page 62

1. Solid should have at least 3 rows of atoms, all touching, with no gaps **[1]**. Liquid should have approximately 11 atoms, at least 50% of them touching, with gaps present **[1]**. Gas should have no atoms touching and show a random arrangement **[1]**.
2. a) Sally would see blue air freshener appearing. [1]
   b) The air freshener is condensing back into a solid from a gas as it cools. [1]
3. Brownian motion is the process by which large particles are hit by the random collision of particles **[1]** causing the large particles to move in random directions **[1]**.
4. When liquid, water molecules can fit closer together **[1]** than when they are arranged as a solid lattice structure **[1]**. This means that the ice is less dense than the liquid form of water **[1]**.
5. b) The CO_2 gas particles are slowed down rapidly causing a solid to form.

Page 63

1. a) [2]

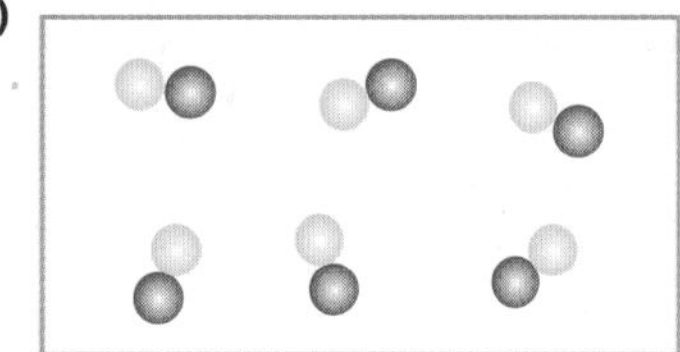

   b) copper oxide [1]
2. c [1]
3. metal + acid → **salt + hydrogen** [1]
4. a) red [1]
   b) orange / yellow [1]
   c) green [1]
   d) blue / purple [1]
5. a) magnesium chloride + hydrogen [1]
   b) copper nitrate + water [1]
   c) vanadium sulfate + water + carbon dioxide [1]

Pages 64–71 **Revise Questions**

Page 65 Quick Test
1. no
2. aluminium
3. anything above sodium in the reactivity series
4. The reactivity series is a list of the comparative reactivity of metals and carbon. It can be used to decide whether a displacement reaction will take place or not.

Page 67 Quick Test
1. They are unreactive, so when found would have been in their elemental form.
2. cool the clay very quickly
3. A combination of materials that, when bonded together, have new, enhanced properties.
4. chemically join monomer units together

Page 69 Quick Test
1. igneous, metamorphic and sedimentary
2. If the chemicals in the lithosphere were soluble, the rain would have washed them away millions of years ago.
3. Check your answer against the picture on p. 69.

Page 71 Quick Test
1. human activity
2. Rare or valuable parts of a product are extracted from waste and then used in a new product or process.
3.

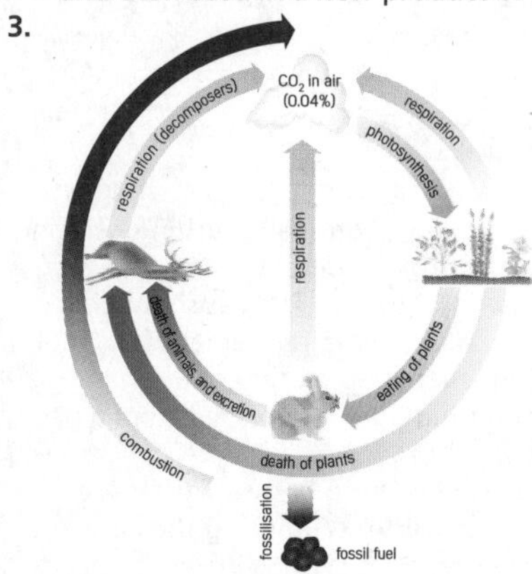

4. Scientists believe that humans are causing climate change through activities such as burning fossil fuels, car use, cutting down rainforests, livestock farming, and paddy fields. The carbon dioxide released is causing the climate to heat up (the greenhouse effect).

Pages 72–73 **Review Questions**

Page 72
1. **a)** A **[1]**
   b) B **[1]**
   c) C **[1]**
2. **a)** The water molecules can get closer together when a liquid compared to when they are a solid **[1]**. This means that liquid water is denser than ice **[1]**. The liquid water at the bottom of the pond will not freeze **[1]**.
   b) If Hilary switches the pump back on, the water will circulate **[1]** and so the water at the bottom will get colder **[1]** and possibly freeze **[1]**.

Page 73
1. **a)** sodium + chlorine → sodium chloride,
   [1 mark for reactants, 1 mark for products]
   b) $2Na(s) + Cl_2(g) \rightarrow 2NaCl(s)$ **[2]**
2. **a)** hydrogen peroxide → water + oxygen **[2]**
   b) a catalyst **[1]**

3. **a)**

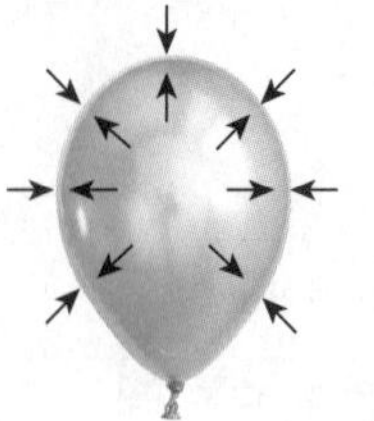

[2]

   b) Before heating – a number of particles drawn, none touching. After heating – same number of particles drawn, further apart! **[1]**

Pages 74–75 **Practice Questions**

Page 74
1. **a)** iron and tin **[2]**
   b) carbon + tin carbonate → **tin** + **carbon dioxide** **[2]**
2. monomers **[1]**
3. Basalt is an igneous rock that cooled very quickly **[1]**. This led to small crystals rather than large **[1]**.
4. A composite is a substance that consists of different materials bonded together **[1]**. Has different properties to the individual substances alone. **[1]**

Page 75
1.

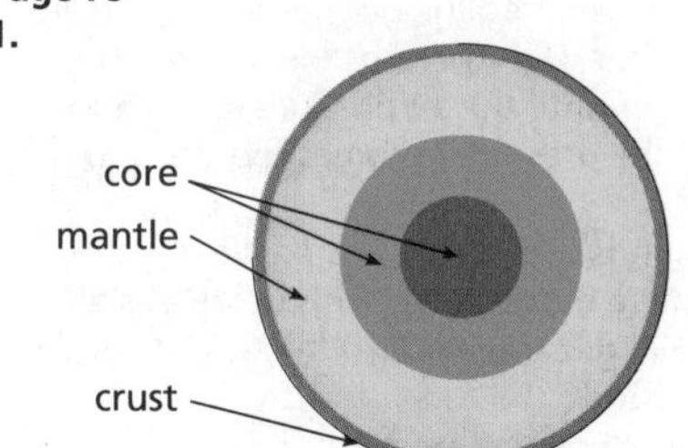

[3]

2.

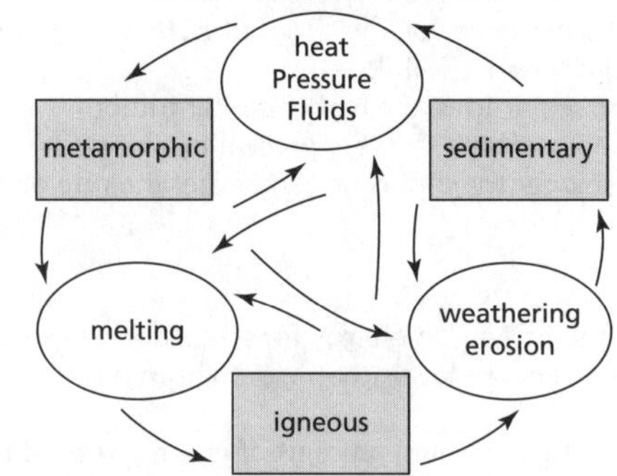

[5]

3. **a)** unusual changes in normal weather patterns **[1]**
   b) ii and iv **[2]**
4. Recycling resources is important as the resources are in limited supply **[1]**. Recycling means that the resources are available for use in other products **[1]**.

Pages 76–83 **Revise Questions**

Page 77 Quick Test
1. unbalanced
2. the Newton (N)
3. Force arrows indicate the direction and strength of a force.
4. force x distance in a given direction; or moment is the turning effect of the force

Page 79 Quick Test
1. m/s or Km/s
2. Work is when an object changes speed or shape.
3. 1.2 km/min or 0.02 km/s or 72 km/h or 20 m/s

Page 81 Quick Test

1. Mass is the amount of particles that make up an object (measured in kg). Weight is a force on the object due to gravity (measured in N).
2. The Earth has gravitational field strength due to its mass. The further away from the centre of mass, the less it will affect an object.
3. An electric field.
4.

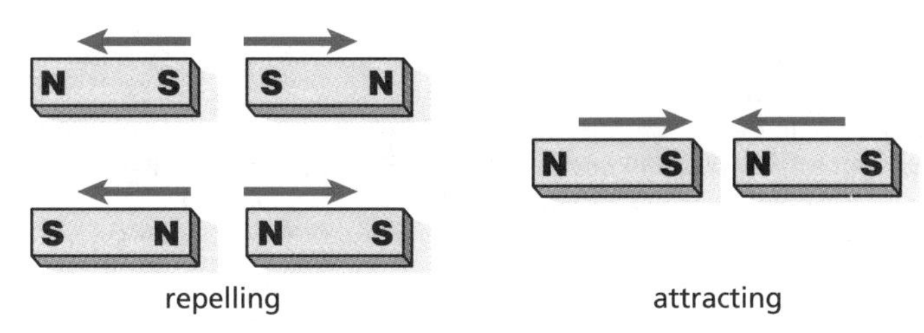

Page 83 Quick Test

1. Atmospheric pressure is due to the weight of air molecules pushing down on the surface. The higher you go, the fewer air molecules there are, therefore less to push down and so less pressure.
2. N/m^2
3. 1000 N
4. The upthrust from the water.

Pages 84–85 Review Questions

Page 84

1. gold **[1]**
2. **a)** ii; iv **[2]**
   b) ii) iron + copper sulfate ➔ iron sulfate + copper **[1]**
   iv) copper + silver nitrate ➔ copper nitrate + silver **[1]**
3. Iron ore is heated at a high temperature with carbon **[1]**. The carbon displaces the iron, which pours out of the bottom of the furnace **[1]**. The carbon reacts with oxygen to form waste carbon dioxide **[1]**.

Page 85

1.

Rock Type	Description
Igneous	is very hard and made of lots of small crystals
Sedimentary	has layers of crystals
Metamorphic	has veins, has a waxy appearance, and is often used for statues

[3]

2.

Element	Abundance (%)
Oxygen	46
Silicon	28
Aluminium	0.8
Iron	0.5

[4]

3. **a)** Arwen and Ruth **[1]**
   b) Recycling does use energy **[1]**. However, the materials are rare and there would be a greater cost in trying to find more and extracting them **[1]**. It is really a question of using less energy **[1]**. **[2 of the 3 possible marks]**

Pages 86–87 Practice Questions

Page 86

1. B (8 N) **[1]**
2. **a)** 200 g **[1]**
   b) 150 g **[1]**
3. 60 s + 48 s = 108 s,
   800 m/108 s = 7.4 m/s **[2]**
4. 15 N x 10 cm = 150 Ncm or 15 N x 0.1 m =1.5 Nm **[3]**

Page 87

1. **a)** S N N S or N S S N **[2]**
   b) Turn one of the magnets the other way round **[1]** so the poles are opposite to one another **[1].**
2. 10 kg – 100 N, 15.5 kg – 155 N, 2000 g – 20 N **[3]**

Watch out for the units. The last mass is in g not kg.

3. **a)**

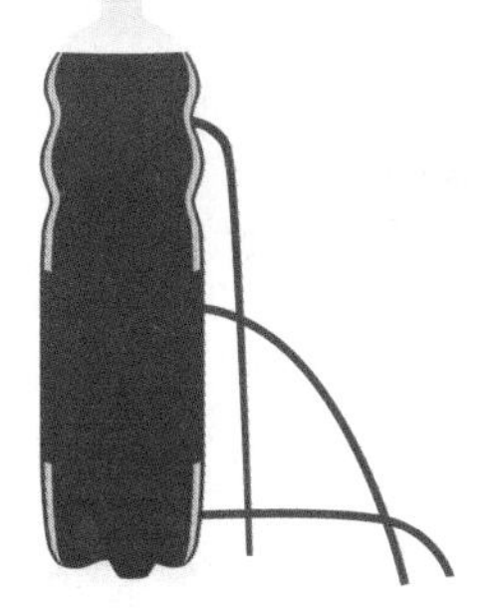

[1]

   b) The pressure on the water is greatest for the lowest hole. The pressure decreases towards the top of the bottle **[1]**, so B will have less pressure than C, but more than A. A has the least pressure **[1]**.
4. Amy 2.4 N/m
   George 1.7 N/m **[2]**

Pages 88–95 Revise Questions

Page 89 Quick Test

1. 1 N upwards
2. A distance–time graph can tell us how far an object travelled over a set time. This also enables speed to be calculated.
3. 130 mph
4. Whether the object is accelerating or decelerating slowly or rapidly.

Page 91 Quick Test

1. 12 h
2. because the Earth is tilted on its axis so the hemispheres receive different amounts of radiation from the Sun.
3. 2000 N
4. Mercury must be smaller in mass and so have lower gravitational field strength.

Page 93 Quick Test

1. fuel + oxygen ➔ carbon dioxide + water + energy
2. Combustion is a much faster reaction, with the energy released producing heat and/or light. Respiration is very slow in comparison.
3. Radiation is the transfer of heat via waves (infrared).
4. 1000 J

Page 95 Quick Test

1. 50 Hz is 50 waves per second
2. An echo is a reflected sound wave.
3. To reduce echoes, place angled materials (like egg boxes) on the walls so that the waves from a singer or instrument are not reflected back.
4. A loudspeaker converts music to an electrical signal which causes electromagnets to move a fabric skin. This movement makes air particles move as sound waves which then reach the ear.

Pages 96–97 Review Questions

Page 96

1. **a)** no **[1]**
   b) yes **[1]**
   c) no **[1]**
2. 600 g – stretch was 12 cm; **[1]**
   700 g – stretch was 14 cm **[1]**
3. Speed = distance ÷ time
   27600 km/h = 42927 km ÷ time **[1]**
   time = 42927 km ÷ 27600 km/h **[1]**
   = 1.56 h (or 93.3 min) **[1]**

Page 97

1. c and d [2]
2. **a)** The rod has been charged with electrons / had electrons removed by rubbing **[1]** and the water, which must have the opposite charge, is attracted to it **[1]**.
   b) it will be repelled (as it has the same charge). [1]
3. The ship does not completely sink **[1]** as, although it is submerged under water, the downward force of the weight **[1]** is counterbalanced by the upthrust or reaction force of the water **[1]**.

Pages 98–99 **Practice Questions**

Page 98

1. A – Lisa jogged for 4 min; B – Lisa stopped for 2 min; C – Lisa walked for 8 min; D – Lisa stopped for 3 min [4]
2. 80 km/h [1]
3. **a)** an astronomical unit of distance, the distance that light travels in an Earth year [1]
   b) 1 light year = 300,000 km/s x 60 x 60 x 24 x 365
   = 9,467,000,000 km [1]
   4.2 light years = 9,467,000,000 km x 4.2 [1]
   = 39,735,360,000 km [1]
   Accept answer rounded to 40,000,000,000 km
4.

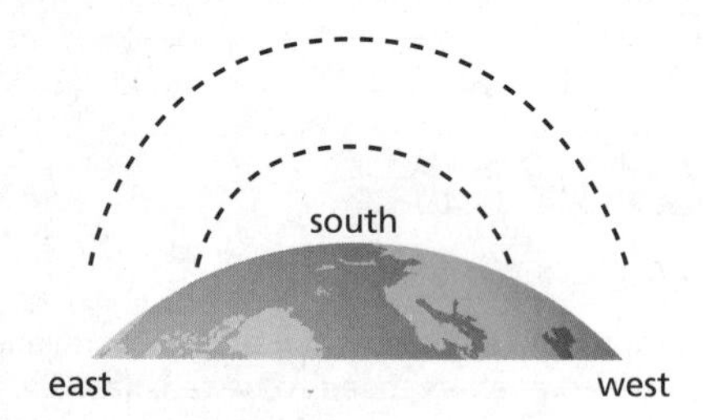

[2]

Page 99

1. **a)** Just when cord goes taut. [1]
   b) As he steps off the bridge, but before he starts falling. [1]
2. **a)** A = oxygen; B = carbon dioxide;
   C = water **[2 marks for all 3 correct]**
   b) respiration [1]
3. c) radiation [1]
4. Jamie and Linda can hear each other because the sound waves are reflected **[1]** by the walls and also spread in all directions from their voices **[1]**.

Pages 100–107 **Revise Questions**

Page 101 Quick Test

1. R = V/I so R = 1.5 V/3 A = 0.5 Ω
2. the ohm (Ω)
3. An insulator is a substance that offers complete resistance to electric current. It does not conduct electricity.
4. **Series circuit**

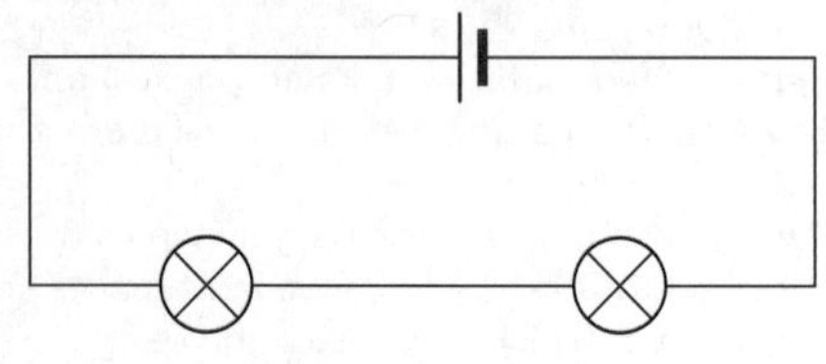

Parallel circuit

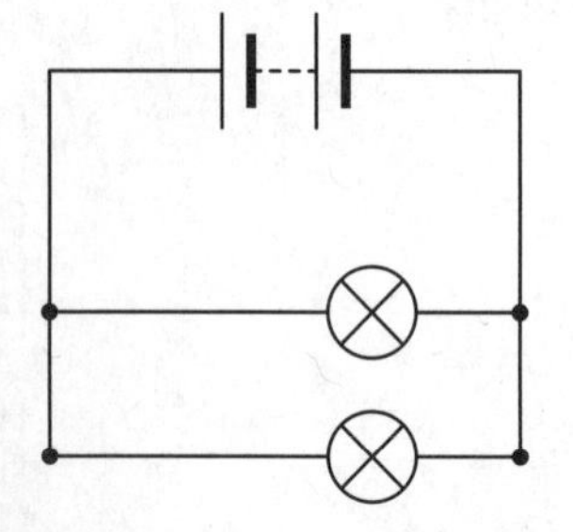

Page 103 Quick Test

1. increase number of coils, increase current, add an iron core
2. The Earth is like a bar magnet because it creates a North and a South pole and the field lines around the Earth resemble those of a bar magnet.
3. The closer the magnetic field lines, the stronger the magnetic field.
4. cobalt, iron and nickel

Page 105 Quick Test

1. light waves do not need a medium to travel through, so they can travel much faster
2. a ripple tank
3. light = 300,000,000 m/s and sound = 300 m/s so light travels approximately 100,000 times faster
4. Waves in phase add together, so increase in size (amplitude).

Page 107 Quick Test

1. violet
2. a prism
3. Light enters into the pinhole and forms an upside down image on the back of the camera. This is where the film would be placed if the image needed to be recorded.
4. where the light hits an uneven surface and reflects in different directions

Pages 108–109 **Review Questions**

Page 108

1. d) travelling at 60 km/h [1]
2. **a)** gravity / gravitational pull from the Sun [1]
   b) 300,000 km/s x 60 x 60 = 108,000,000 km/h
   4.1 h x 108,000,000 km/h = 4,428,000,000 km [3]
3. The Earth has seasons because the planet is tilted **[1]** in respect to the North Star. This means that, as the Earth goes around the Sun the summer is when a hemisphere is tilted towards the Sun **[1]** and winter is when a hemisphere is tilted away from the Sun **[1]**.

Page 109

1. **a)** i and iv [2]
   b) iv [1]
2. **a)** As we get older we lose our ability to hear higher pitched/ high frequency sounds **[1]**. Hence teenagers can hear the noise, but the adults cannot **[1]**.
   b) The loudspeaker vibrates **[1]**, hitting air particles which gain energy **[1]**. They move in waves to the teenager's ear **[1]**. The eardrum vibrates at the same frequency as the original sound **[1]**.
3. b) The bomber is a good reflector of sound. [1]

Pages 110–111 **Practice Questions**

Page 110

1.

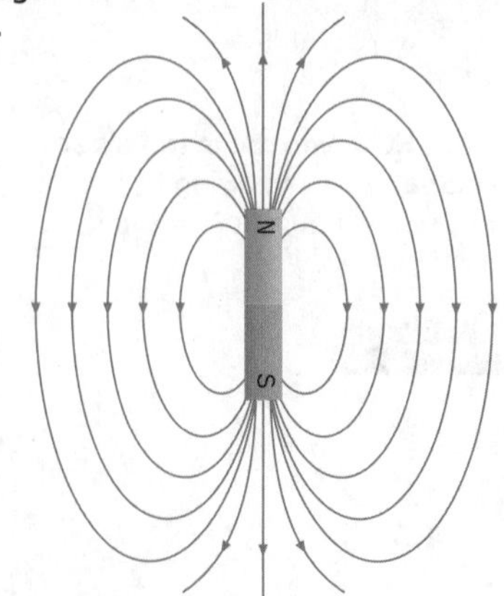

[2]

2. **a)** i; iv; vi [3]
   b) the electromagnet can be turned on and off [1]

3. A2 and A3 = 2 A, A4 = 4 A **[2]**
4. **a)** P = I x V so I = P/V =
   60 W / 230 V = 0.26 A **[1]**
   b) 60 x 200 = 12,000 Wh = 12 kWh **[3]**

Page 111

1.

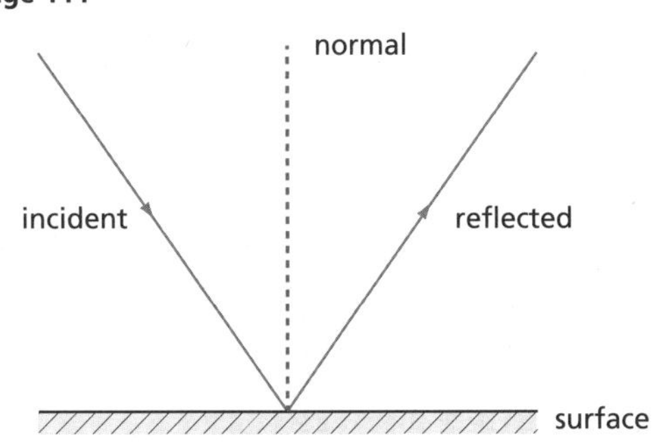

[3]

2. **a)**

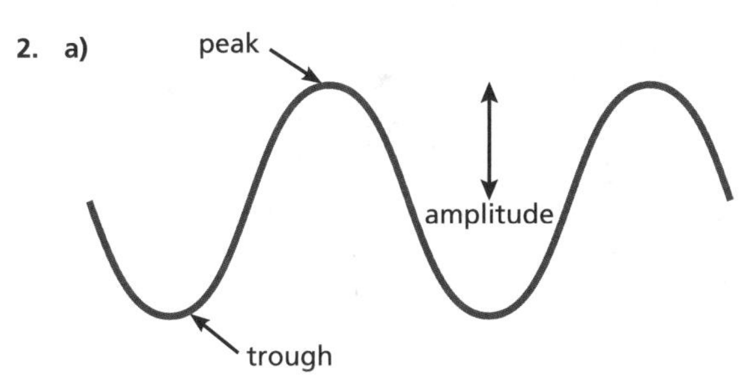

[3]

   b) 2 Hz **[2]**
3. red, orange, yellow, green, blue, indigo, violet (top to bottom on diagram) **[4]**

Pages 112–113 **Review Questions**

Page 112

1. c) cobalt **[1]**
2. C **[1]**
3. **a)** it will turn on/activate and attract the armature **[1]**
   b) it will be attracted to the electromagnet **[1]** and complete the second circuit **[1]**

Page 113

1. In a mirror, all the waves are reflected at the same angle **[1]**. This is specular reflection **[1]**. With paper, the rays hit the uneven surface and reflect in different directions. This is called scattered diffusion **[1]**.
2. **a)** 10 Hz means 10 waves per second **[1]**
   b) Four added waves **[1]** same wavelength **[1]** (See diagram below)
   c) Four added waves reflected from the bar **[1]** same wavelength as incoming waves **[1]** (See diagram below)

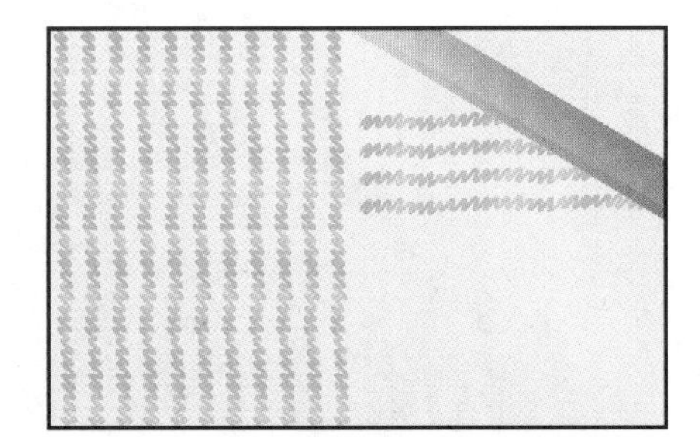

3. **a)** **[3]**

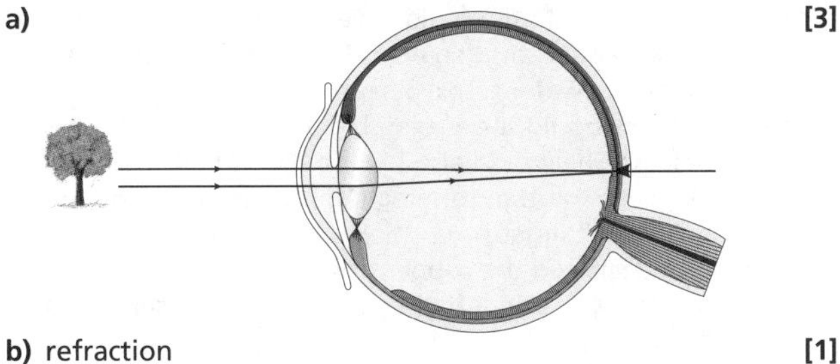

   b) refraction **[1]**

Pages 114–125 **Mixed Test-Style Questions**

Page 114

1. **a) i)** B should be drawn at the top of the apparatus, inside the round bottomed flask. **[1]**
   ii) The distillate is the liquid collected at the end of the distillation tube. The label X should appear in the liquid, or be clearly labelled with a line to the collected liquid. **[1]**
   b) The cold water enters at the bottom of the distillation tube and leaves at the top **[1]**. This is so that the gas will always encounter colder conditions as it moves down the distillation tube. This means it will condense more rapidly. **[1]**
   c) 9 **[1]**
   d) Point Y is a liquid. At least 11 particles need to be drawn with at least 50% of the particles touching; Point Z is a gas. A minimum of five particles need to be drawn with no particles touching **[both diagrams need to be drawn correctly for 1 mark].**
2. **a)** 20 N **[1]**
   b) Michael is already using 20 N of force.
   Kerrie weighs 600 N, so 600 N / 2 = 300 N.
   Therefore 320 N of force used. **[2]**
   If part (a) is incorrect and in part (b) only uses 600 N, follow through and accept 300 N (with working).
   If part (a) is correct but in part (b) only uses 600 N, award 1 mark for working.
   c) i) pressure = force / area
   Force = 650 N. The area of 1 stilt is 9 cm^2. There are 2 stilts.
   Therefore total area = 2 x 9 cm^2 = 18 cm^2.
   650 N / (2×0.09), 650/0.18 = 3611 N/cm^2
   [1 mark for correct working, 1 mark for correct answer]
   If working shows only 9 cm^2 used, award maximum 1 mark.
   Also accept 650 N/(2 x 0.18 m) = 3611 N/m^2
   ii) To exert less pressure she needs to use stilts with a larger surface area or lose weight. **[1]**
3. **a)** Breathing in is shown by the rubber sheet being pulled down **[1]**; volume in the bottle increases **[1]** and pressure decreases so air enters the balloons **[1]**. Breathing out is shown by the rubber sheet going up **[1]**, volume decreases **[1]**, pressure increases and forces air out of the balloons **[1]**.
   b) i) asthma **[1]**
   ii) smoker **[1]**
   iii) normal **[1]**
   c) i) 10→8; 10→6; 8→6 **[3]**

> Diffusion always moves from high to low concentration.

   ii) Glucose moves by diffusion **[1]** from a high concentration **[1]** to a low concentration **[1]**.
   d) i) anaerobic makes lactic acid or aerobic doesn't make lactic acid **[1]**;
   aerobic releases more energy or anaerobic releases less energy **[1]**
   ii) humans produce lactic acid **[1]**; yeast produces alcohol **[1]**

> Remember *aerobic* uses oxygen in *air*.

4. **a)** A new material has been deposited on the surface of the nail. **[1]**
   b) iron + oxygen + water **[1]** → iron hydroxide **[1]**

c) Naveen's hypothesis would have been: an iron nail will only rust if water and oxygen are present / an iron nail will not rust if either water or oxygen are missing. **[2]**

d) i) Naveen would see a reddish–brown metal forming on the nail **[1]**. She would also see the blue colour disappearing / a green solution forming **[1]**. [**Accept the colour of the nail has changed**]

ii) iron displaced the copper from the sulfate and takes its place **[1]**; iron is higher in the reactivity series than copper **[1]**

5. a) Numbers of blue tits reduce **[1]** because fewer insects would be available to eat **[1]**. Numbers of moths increase **[1]** because there is less competition for their food from the leaf-eating insects **[1]**.

This question is about interdependence, i.e. how one organism in a food web affects another organism.

b)

	✓ or ✗
Respiration stores energy as light	✗
Photosynthesis uses energy from light	✓
Respiration breaks down large molecules to smaller molecules	✓
Photosynthesis creates organic molecules from inorganic molecules	✓
Respiration releases energy from chlorophyll	✗
Photosynthesis stores energy as light	✗

[6]

c) i) extinction means all members of a species have died out worldwide **[1]**

Remember, once an organism is extinct it cannot return.

ii) a change in their environment **[1]** which happened rapidly **[1]**

d)

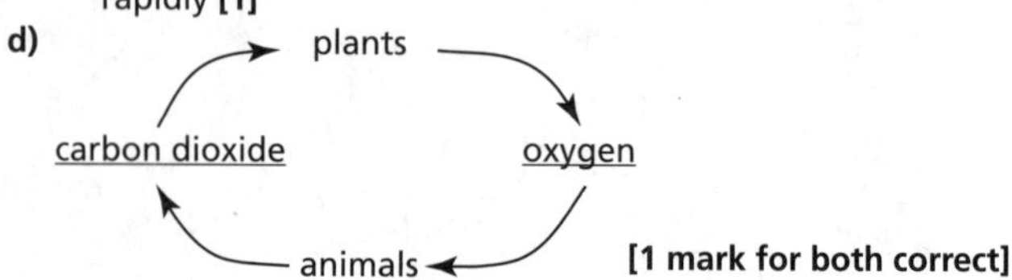

[1 mark for both correct]

e) Most scientific discoveries are the result of a scientist building on the work of a previous scientist **[1]**. Scientists have their results checked by other scientists **[1]**. Scientists use data from experiments to check their ideas **[1]**.

6. a) i) Carriage X should have S–N; carriage Z should show N–S. **OR** carriage X should have N–S; carriage Z should show S–N. **All poles need to be correct, none missing. [1]**

ii) Carriage X would now be attracted to carriage Y, whereas carriage Z would repel carriage Y **[1]** this is because opposite poles attract and like poles repel **[1]**.

b) A, steel; B, magnet; C, aluminium **[2 marks for all 3 correct, 1 mark for 2 correct, no marks for only 1 correct].**

c) cobalt and iron should be ticked **[1]**